Electron Microscopy and Analysis Group Conference (EMAG 2015)

Journal of Physics: Conference Series Volume 644

Manchester, United Kingdom
29 June – 2 July 2015

ISBN: 978-1-5108-1471-4
ISSN: 1742-6588

Printed from e-media with permission by:

Curran Associates, Inc.
57 Morehouse Lane
Red Hook, NY 12571

Printed by Curran Associates, Inc. (2016)

For permission requests, please contact the Institute of Physics at the address below.

Institute of Physics
Dirac House, Temple Back
Bristol BS1 6BE UK

Phone: 44 1 17 929 7481
Fax: 44 1 17 920 0979

techtracking@iop.org

Additional copies of this publication are available from:

Curran Associates, Inc.
57 Morehouse Lane
Red Hook, NY 12571 USA
Phone: 845-758-0400
Fax: 845-758-2634
Email: curran@proceedings.com
Web: www.proceedings.com

Table of contents

Volume 644

Electron Microscopy and Analysis Group Conference (EMAG2015)
29 June to 2 July 2015, Manchester, UK

Papers

Functional Materials

Interfaces and Coatings

3D Microscopy in SEM and TEM

FIB and SEM

Nanomaterials

EM of Magnetic and Structural Materials

Advanced EM Techniques

EM of Biological Systems and Biomaterials

EMAG2015

2015 marked a new venture for the EMAG group of the Institute of Physics in that the conference was held in conjunction with the MMC2015 conference at the wonderful Manchester Central conference centre. As anyone who was there would be able to confirm, this went exceptionally well and was a really vibrant and top quality conference. The oral sessions were filled with good talks, the poster sessions were very lively, and there was a good balance between oral sessions with a specifically "EMAG" identity, and the integration into a larger conference with the ability to switch between up to six parallel sessions covering physical sciences, techniques, and life sciences. The large conference also attracted a wide range of exhibitors, and this is essential for the ongoing success of all of our work, in a field that is very dependent on continued technical innovation and on collaborations between academic researchers and commercial developers of microscopes, holders, detectors, spectrometers, sample preparation equipment, and software, among other things.

As has long been the case at EMAG, all oral and poster presenters were invited to submit papers for consideration for the proceedings. As ever, these papers were independently reviewed by other conference attendees, with the aim of continuing the long tradition of the EMAG proceedings being a top quality, peer-reviewed publication, worthy of reference in future years. Whilst I recognise that not all presenters were able to submit papers to the proceedings (for instance due to the need not to prejudice publication in some other journals, or due to avoiding duplicate publication of data), we are gratified that our presenters submitted as many papers as they did. The 41 papers included provide an interesting snapshot of many of the areas covered in the conference presentations, including functional materials, coatings, 3D microscopy, FIB and SEM, nanomaterials, magnetic and structural materials, advances in EM techniques, and EM of biological systems and biomaterials. The arrangement of the papers in the proceedings is not the same as the ordering of the conference sessions, but has been grouped by similar topics. I hope you enjoy reading these papers and that these serve as an interesting summary of what was an excellent conference.

Best wishes

Dr Ian MacLaren
SUPA School of Physics and Astronomy
University of Glasgow
Glasgow G12 8QQ

Peer review statement

All papers published in this volume of *Journal of Physics: Conference Series* have been peer reviewed through processes administered by the proceedings Editors. Reviews were conducted by expert referees to the professional and scientific standards expected of a proceedings journal published by IOP Publishing.

Investigating the effects of contacting functional oxide films

K J O'Shea[1], N Homonnay[2], G Schmidt[2] and D A MacLaren[1,3]
[1]SUPA School of Physics and Astronomy, University of Glasgow, G12 8QQ, UK
[2]Fachbereich Physik, Martin-Luther-Universität Halle-Wittenberg, Halle (Saale), Germany

E-mail: dmaclaren@physics.org

Abstract. Advanced oxides are the current focus of intense research activity, driven by their numerous attractive properties for next generation microelectronics. However, reliable strategies must be developed for the electrical contacting of such films without compromising their functionality. We explore the effect of depositing both noble and oxidising metals onto the ferromagnetic $La_{0.7}Sr_{0.3}MnO_3$ (LSMO) in terms of its structural and magnetic properties. Whilst noble metal overlayers have negligible impact, the metals typically used as adhesion layers, such as Ti, can drive a structural phase transition in the LSMO, producing a Brownmillerite phase and impairing the magnetisation.

1. Introduction

Functional oxides are promising candidates for next generation microelectronics owing to their wide range of ferroelectric, ferromagnetic and potential emergent interfacial properties such as interfacial superconductivity (within heterostructures) [1]. In addition to room temperature spintronic applications, functional oxides are also highly valuable in the investigation of spin dynamics [2] where the principal focus lies at the interface between ferromagnetic oxides and metals. Critically, all metal-oxide systems rely on smooth interfaces, good electrical contact and a lack of interfacial chemical reactions, therefore a complete understanding of the metallisation of functional oxides is essential for metal-oxide hybrid devices to be fully realised.

Here, we present a systematic study of the magnetic and structural effects of depositing metal contacts onto $La_{0.7}Sr_{0.3}MnO_3$ (LSMO) thin films using transmission electron microscopy (TEM) and superconducting quantum interference device (SQUID) magnetometry. We find that complete quenching of the magnetisation of >10 nm oxide can occur after the deposition of just ~1 nm of a reducing metal. This degradation is important in the context of forming reliable metal contacts to functional oxides, especially where adhesion layers such as titanium are conventionally used.

2. Experiment Details

Pulsed laser deposition (PLD) [3] was used for epitaxial growth of LSMO, using (100)-oriented $SrTiO_3$ (STO) substrates and stoichiometric polycrystalline targets. The PLD system employed a KrF excimer (248 nm) laser and the growth was monitored in-situ by high pressure reflection high energy electron diffraction (RHEED) [4]. Ten nanometre and 20 nm thick LSMO films were deposited using

[3] To whom any correspondence should be addressed.

a laser fluence of 2 - 3 J/cm^2, a repetition rate of 5 Hz, an O_2 atmosphere of 0.2 mbar and a substrate temperature of 700°C. Following deposition of the oxide layer(s), samples were transferred to the electron beam evaporation chamber under UHV for deposition of the metals.

Scanning transmission electron microscopy (STEM) was used for structural characterisation of the metallised films and was performed on a JEOL ARM 200cF instrument. Electron energy loss spectroscopy (EELS) was used for chemical analysis of the interfaces and was conducted using a Gatan Quantum 965 spectrometer, typically using a dispersion of 0.5 eV/channel.

3. Results

Figure 1 provides an overview of the effect of depositing Au, Pt, Cu, CoFe, Cr and Ta electrical contacts on the crystal structure of LSMO films. Figures 1a and b are high angle annular dark field (HAADF) images of an LSMO film with a gold and platinum capping layer, respectively, with the Au/Pt layer appearing brightest. There are no discernible effects on the LSMO crystal structure resulting from the deposition of either Au or Pt (figures 1a,b), which is perhaps unsurprising given that neither of these elements is particularly reactive. SQUID measurements (figure 1g) indicate no substantial reduction in the magnetic moment of the LSMO with either a gold or platinum cap. In contrast, the SQUID measurements indicate a slight reduction in magnetic moment for a Cu contact and the LSMO crystal structure appears more diffuse (figure 1c), with a dark narrow band of intensity appearing at the Cu/LSMO interface in HAADF imaging. Similarly, the deposition of both Cr and Ta (figures 1e,f) produce a more prominent dark interfacial band and an almost complete loss of magnetisation. In each case, the dark band is consistent with the formation of an interfacial oxide, with the prospect of yielding an electrically insulating barrier layer between metal and LSMO.

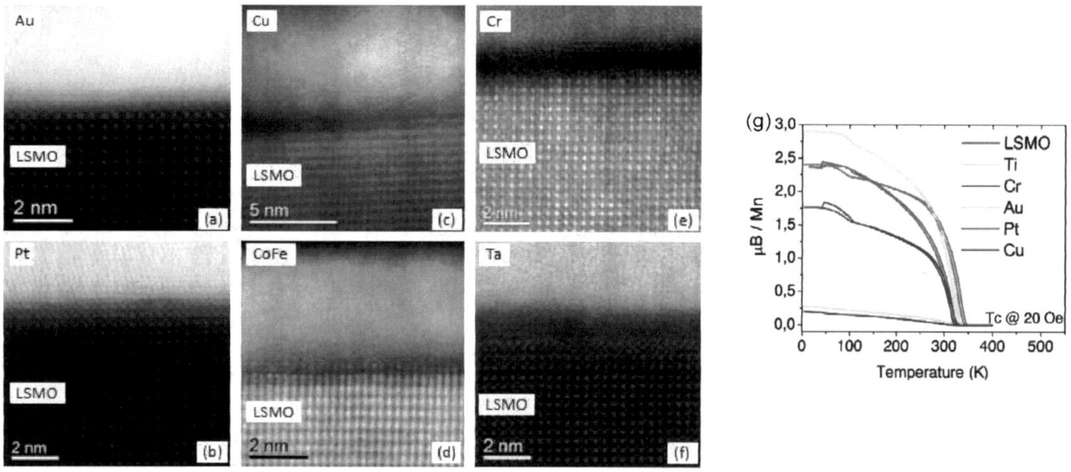

Figure 1.High angle annular dark field (HAADF) image summary of the effect of different metal contacts on LSMO films: (a) Au, (b) Pt, (c) Cu, (d) CoFe, (e) Cr and (f) Ta. There are no discernible effects for Pt and Au, however a significant interface region forms in the cases of Cu, CoFe, Cr and Ta, all of which are commonly used as electrical contacts or (for CoFe) magnetic contacts. SQUID measurements (f) indicate a substantial reduction in magnetisation following the deposition of Ti and Cr compared to the uncoated LSMO (red trace), whilst Cu reduces the magnetisation slightly and Au and Pt have no detrimental effect. The apparent increase in magnetization upon Au deposition is ascribed to sample-to-sample variations in LSMO quality.

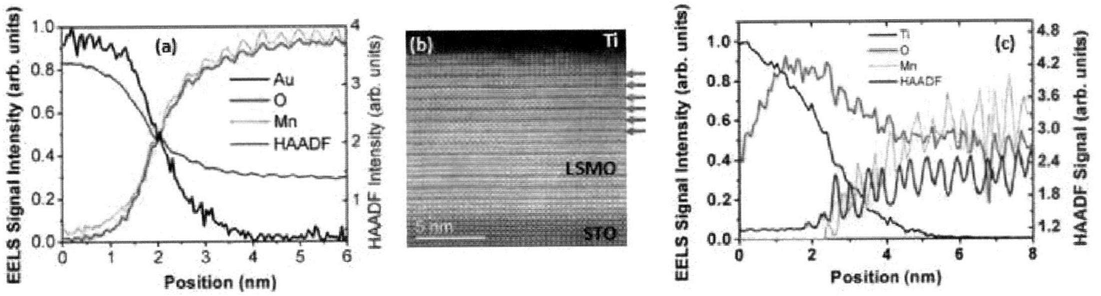

Figure 2. (a) EELS signal profiles of the Au (black), O (red) and Mn (green) across the LSMO/Au interface given in figure 1a, acquired using the Dual EELS [5] and spectrum imaging [6] methodologies within Digital Micrograph™ software. The HAADF signal intensity (blue) is also included for comparison. (b) HAADF image of the effect of Ti deposition onto LSMO. Red arrows highlight dark lattice planes which appear at a periodicity of twice the unit cell. (c) EELS signal profiles of the Ti (black), O (red) and Mn (green) across the LSMO/Ti interface. The HAADF signal highlights the cell doubling effect.

Figure 2a shows the EELS signal intensities of the Au $M_{4,5}$, O K and Mn $L_{2,3}$ edges across the LSMO/Au interface, in addition to the HAADF signal intensity (blue) for comparison. A small degree of intermixing is apparent at the interface but is ascribed to the data deriving from the projection through a rough interface, in addition to possible beam-broadening effects. Whilst the LSMO is unaffected by the deposition of a gold contact, a dramatic effect is observed following the deposition of Ti and explains the quenched magnetisation indicated in figure 2g. A HAADF image is given in figure 2b where the LSMO now appears bright with respect to the Ti. A clear cell-doubling effect is observed in the LSMO layer, which is consistent with a phase change to a brownmillerite structure [7,8]. The dark planes in figure 2b, highlighted by the red arrows, coincide with the MnO_2 planes of the LSMO, indicating a deficiency in either manganese or oxygen. EELS was subsequently carried out to determine the origin of the phase change in the LSMO and results are summarised in figure 2c. It is immediately clear that the oxygen signal extends some distance into the Ti capping layer, decaying at a slower rate than the Mn at the interface. This indicates that the Ti acts as an oxygen getter and subsequently forms a TiO_x phase at the interface, consistent with an analysis of the Ti $L_{2,3}$ ELNES spectrum (not shown) [9]. These results can account for the complete loss of magnetisation in the LSMO since the sensitivity of LSMO magnetisation to structural defects and oxygen deficiency is well known [10].

In the next set of experiments, we explore the effect of Ti thickness on the magnetic properties of LSMO. Figure 3a illustrates that complete quenching of the LSMO magnetisation occurs for just 1 nm of Ti. What is interesting is that the crystal structure of the LSMO appears completely intact, as illustrated in figure 3b. Nevertheless, EELS reveals substantial oxygen deficiencies,

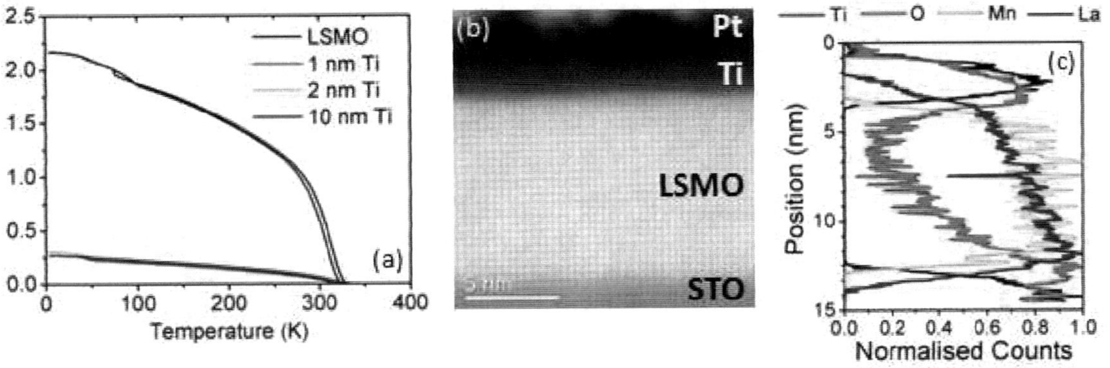

Figure 3. (a) SQUID measurements of Ti capping layers of different thicknesses. (b) STEM image of an LSMO film with a 1 nm thick Ti cap. (c) Corresponding EELS profiles of Ti, O, Mn and La signals across the film in (b) where the LSMO is oxygen (red) depleted.

indicating that the LSMO is able to support a significant number of oxygen vacancies prior to reconstruction. The data presented in figure 3c show the Ti $L_{2,3}$, O K, Mn $L_{2,3}$ and La $M_{4,5}$ intensity profiles across the LSMO layer following the deposition of ~1 nm of Ti. It may be observed that there is a significant oxygen signal (red line) in the Ti layer, whilst the LSMO layer is substantially oxygen depleted, suggesting an interfacial reaction and transfer of oxygen from LSMO to Ti, with the oxygen deficiency becoming more pronounced towards the Ti interface.

4. Discussion and Conclusions
Given that Ti and Cr are often used as adhesion layers for noble metal electrical contacts, these results indicate that great care must be taken to protect oxygen sensitive materials from damage. Whilst an interfacial reaction is commonly exploited to improve the adhesion of ohmic contacts in microelectronic devices, it is clear that in the case of functional oxides such as LSMO, that reaction can have a detrimental effect on functionality. A suitable candidate for use as a protective barrier sandwiched between LSMO and Ti was found to be the electrically conductive perovskite oxide $SrRuO_3$. Alternatively, we found that $SrTiO_3$ could also be used as a suitable mask for subsequent dry etching if an ohmic contact is not required; these results will be described elsewhere. The results presented here provide valuable insight into the effects of metal contacting and ultimately enhance our understanding towards the development of multifunctional oxide devices.

Acknowledgements
This work was funded by EU's 7th framework Program IFOX (NMP3-LA-2010 246102) and the Engineering and Physical Sciences research council (EPSRC) of the UK (EP/I00419X/1). Original data files relating the work presented are available at: http://dx.doi.org/10.5525/gla.researchdata.193.

5. References
[1] Hwang H Y, Iwasa Y, Kawasaki M, Keimer B, Nagaosa N and Y. Tokura 2012 *Nature Mater.* **11** 103
[2] Czeschka F D, Dreher L, Brandt M S, Weiler M, Althammer M, Imort I-M, Reiss G, Thomas A, Schoch W, Limmer W, Huebl H, Gross R and Goennenwein S T B 2011 *Phys. Rev. Lett.* **107** 046601
[3] Willmott P R and Huber J R 2000 *Rev. Mod. Phys.* **71** 315
[4] Neave J H and Joyce B A, Dobson P J, Norton N 1983 *Appl. Phys.* A **31** 1.
[5] Scott J, Thomas P J, MacKenzie M, McFadzean S, Wilbrink J, Craven A J and Nicholson W A P 2008 *Ultramicroscopy* **108** 1586
[6] Hunt J A and Williams D B 1991 *Ultramicroscopy* **38** 47
[7] Rossell M D, Lebedev O I, Van Tendeloo G, Hayashi N, Terashima T and Takano M 2004 *J. Appl. Phys.* **95** 5145
[8] Klenov D O, Donner W, Foran B, Stemmer S 2003 *Appl. Phys. Lett.* **82** 3427
[9] Stoyanov E, Langenhorst F, Steinle-Neumann G 2007 *American Mineralogist* **92** 577
[10] Schumacher D, Steffen A, Voigt J, Schubert J, Brückel Th, Ambaye H and Lauter V 2013 *Phys. Rev. B* **88** 14427

Structural investigation of epitaxial $LaFeO_3$ thin films on (111) oriented $SrTiO_3$ by transmission electron microscopy

E Christiansen[1], M Nord[1], I Hallsteinsen[2], P E Vullum[1,3], T Tybell[2], R Holmestad[1].

[1] Department of Physics, Norwegian University of Science and Technology (NTNU), 7491 Trondheim, Norway.
[2] Department of Electronics and Telecommunications, Norwegian University of Science and Technology (NTNU), 7491 Trondheim, Norway.
[3] SINTEF, Materials and Chemistry, 7465 Trondheim, Norway.

E-mail: randi.holmestad@ntnu.no

Abstract. We report on structural domains in $LaFeO_3$ epitaxial thin films on (111) oriented $SrTiO_3$ observed by transmission electron microscopy using low magnification dark field imaging and high resolution transmission electron microscopy techniques. The films were grown by pulsed laser deposition and had a thickness $\approx$ 20 nm. Three domain orientations are found, in accordance with the orthorhombic structure of $LaFeO_3$. The domains themselves are of irregular shapes, and vary in size from tens to thousands of nm^2. Regions of reduced Bragg scattering are observed as straight lines along $< 100 >$, hinting at a complex domain structure.

1. Introduction

Antiferromagnetic (AFM) materials play an important role in magnetic device technology, such as the spin valves [1]. In such devices, AFM layers are used to pin adjacent ferromagnetic layers through exchange coupling [2]. In order to apply novel functional magnetic systems in practical applications, such as magnetic data memory, thin films with sharp interfaces between the AFM and ferromagnetic layers are necessary [3]. Perovskite oxide materials are in this regard interesting, exhibiting functional properties including both ferromagnetic and antiferromagnetic systems [4]. Based on their strong structure-property coupling, the functional properties of perovskite thin films can be tailored by synthesis [5].

$LaFeO_3$ (LFO) is a G-type antiferromagnetic perovskite [6]. It belongs to the orthorhombic space group $Pbnm$ (no. 62) and has lattice parameters $a = 5.556$ Å, $b = 5.565$ Å, and $c = 7.862$ Å (pseudo cubic parameters $a_{pc} = 3.929$ Å, $b_{pc} = 3.935$ Å, and $c_{pc} = 3.931$ Å) at room temperature, which makes it belong to the $a^-_{pc}a^-_{pc}c^+_{pc}$ Glazer tilt system [7, 8]. This means that the oxygen octahedra are tilted out-of-phase along a_{pc} and b_{pc}, and in-phase along c_{pc}. Its Neél temperature of $T_N = 740$ K is highest among the orthoferrites, and is attributed to the large buckling angle of the Fe-O-Fe bonds related to the octahedra tilts [9, 10]. For depositing LFO thin films, a common substrate is $SrTiO_3$ (STO), a nonmagnetic and insulating cubic $Pm\bar{3}m$ (no. 221) perovskite with $a_{STO} = 3.905$ Å (parameters and indices without a subscript are LFO indices unless otherwise specified, while subscript *STO* is used to denote STO axes throughout this paper).

For thin films of LFO on (100) and (110) STO, the Néel temperature is slightly reduced [11]. Due to its orthorhombic symmetry, LFO forms structural domains on STO. Moreover, photoelectron emission microscopy data indicate AFM domains of the same shape and size as the structural domains. In the present work, the structure of an epitaxial thin film of LFO on (111) oriented STO grown by pulsed laser deposition (PLD) is studied by transmission electron microscopy (TEM). This in order to study possible effects off the quasi-hexagonal surface symmetry of this facet on structural domain formation. Domains in the film is studied by electron diffraction and dark field (DF) imaging of plan-view samples, and the LFO/STO interface is studied by high resolution TEM (HRTEM) of cross section samples.

2. Materials and methods

The $LaFeO_3$ thin film was synthesised by PLD, using a KrF excimer laser (wavelength $\lambda = 248$ nm) with a fluency of ~ 2 J/cm^2 and repetition rate 1 Hz impinging on a stoichiometric $LaFeO_3$ target. A deposition temperature of 540 °C, oxygen pressure of 0.35 mbar and target-substrate distance of 45 mm was used to ensure quality growth [12]. The substrate was prepared by buffered hydrogen fluoride etch and annealing in oxygen ambient, resulting in a step and terrace structure with miscut of 0.1° and step edges along a $< 11\bar{2} >$ STO direction. Reflection high energy electron diffraction was used to monitor the growth in situ, ensuring layer-by-layer growth. Atomic force microscopy confirmed a high surface quality of the film, and the film was found to be fully epitaxial by x-ray diffraction.

TEM samples of both plan-view and cross section geometries were prepared. Plan-view samples were fabricated by grinding film wafer pieces from the substrate side into a wedge shape (wedge angle of 4°), using an *Allied MultiPrep Polishing System* (tripod). One plan-view sample was further thinned in a *FEI Helios NanoLab DualBeam FIB* in order to remove surface scratches from the tripod polishing, and provide unbent electron transparent regions. Pure tripod plan-view samples were also fabricated to control that the FIB process did not alter the structure of the film. The sample that was Ga^+ ion-beam thinned yielded the same results as samples prepared solely by tripod polishing, demonstrating that ion-beam thinning did not alter the crystallography of the material system. All presented images are from samples that were prepared by low energy (2 kV) ion-beam thinning, as pure mechanical polishing leaves polishing stripes and other surface defects. The same FIB instrument was used to fabricate cross section samples from the same film, using a lift-out procedure.

Dark field imaging was performed on a *JEOL JEM-2100* LaB_6 TEM. In order to better show and compare the relationships between the three different structural domains, several DF images taken with each of the unique orthorhombic reflections were merged into a single image. Domains with equal crystallographic orientations are given the same colour (see Fig. 1). For HRTEM imaging a *JEOL JEM-2100F* field emission gun TEM was used. Both TEMs were operated at 200 kV acceleration voltages.

3. Results and discussion

A domain mosaic, combining six DF images from a plan-view sample, is presented in Fig. 1. The six DF images are taken with each of the six orthorhombic super reflections, located midway between 000 and each of the six $< 1\bar{1}0 >_{STO}$ reflections. Since the TEM sample is a bilayer with the 20 nm thick LFO film on top of the substrate, the electron beam will pass through both the film and the substrate and the resulting diffraction pattern will thus contain information about both layers simultaneously. However, the reflections chosen for each of the dark field images are unique for the LFO film and not present in the diffraction pattern from pure STO. As such, the contrast from elastic Bragg scattering in the DF images will only come from the LFO film.

Furthermore, the chosen DF reflections are the orthorhombic 100 or 010 reflections. This also means that we have at least three different structural domains in the LFO film. Each structural domain contributes to two of the six super reflections. The domain mosaic in Fig. 1 is composed of six DF images, even if three theoretically would be enough, but using all six reflections helps to cancel out contrast changes due to sample bending that will give small variations of the sample orientation.

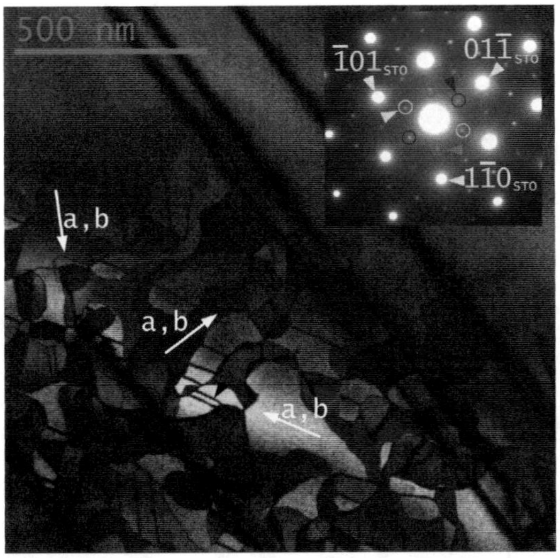

Figure 1. DF mosaic of six individual DF images. The reflections used are circled in the inset electron diffraction pattern, the colour of the circles indicate the colour assigned to each DF image. The arrowheads in the diffraction pattern mark LFO (100)/(010) reflections, and the white arrows in the real space image show the direction of [100]/[010]. Dark lines are seen inside some domains, one case being pointed out by a black arrowhead. Such lines tend to be parallel to [100]/[010].

In Fig. 1 we observe that the orthorhombic [100] or [010] directions can be parallel to $[1\bar{1}0]_{STO}$, $[\bar{1}01]_{STO}$ or $[01\bar{1}]_{STO}$, leading to the three different structural domains with red, green and blue colours in the figure, respectively. However, we are not able to discriminate between the orthorhombic [100] and [010] directions since the orthorhombic a and b lattice parameters are almost equal. This means that two regions with the same colour in Fig. 1 does not necessary have the same orientation of the unit cell, as the orthorhombic a and b lattice parameters can switch between any two regions with the same colour. We also observe lines inside one-coloured regions where we have insignificant Bragg scattering to any of the six super reflections. If these lines are related to for instance twin boundaries where the orthorhombic a and b lattice parameters switch directions across the dark line, is something that future characterization work must answer.

A HRTEM image from a cross section sample with $[11\bar{2}]_{STO}$ zone axis orientation is presented in Fig. 2, revealing a coherent film/substrate interface and high crystalline quality film. The Fourier transforms (inset in the HRTEM image) from the left (region A) and the right (region B) sides of the film clearly show that more than one structural domain are present in the field of view. In region A we have a structural domain where the orthorhombic a or b lattice parameter is parallel to the $[1\bar{1}0]$ direction of STO. In region B the orthorhombic unit cell is oriented differently with none of the orthorhombic lattice parameters being parallel to the $[1\bar{1}0]_{STO}$ direction.

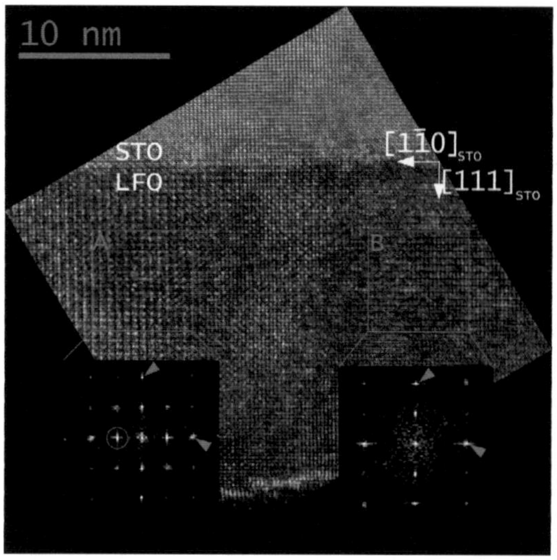

Figure 2. HRTEM image of LFO thin film cross section sample acquired in the $[11\bar{2}]_{STO}$ zone axis. The film is epitaxial, but not structurally homogeneous, as the FFT spectra of region "A" and "B" are clearly different, indicating the presence of at least two different domains. The lowest spatial frequencies of STO have been marked by arrowheads for reference. The circled frequency component in the FFT of region A indicates that either a or b is aligned with the $[\bar{1}10]_{STO}$ direction in this domain. As region B lacks this frequency component, the a or b axis may not be parallel to the $[\bar{1}10]_{STO}$ direction here.

4. Conclusions

Structural domains of irregular shape and size were found in $\approx$ 20 nm thin LFO epitaxial films on (111) STO by DF TEM. The difference between the structural domains is that the orthorhombic a or b lattice parameters can be oriented parallel to either one of the substrate's $[1\bar{1}0]$, $[\bar{1}01]$ or $[0\bar{1}1]$ directions. However, since we are not able to discriminate between the orthorhombic a and b lattice parameters, it is still not clear if the film consists of more than 3 structural domains. However, dark contrast lines inside one-coloured regions, i.e. regions with similar diffraction patterns, indicate that further work is needed to fully understand the complete structure of the LFO film.

References

[1] Wolf S A, Awschalom D D, Buhrman R A, Daughton J M, von Molnár S, Roukes M L, Chtchelkanova A Y and Treger D M 2001 *Science* **294** 1488–95

[2] Nogués J and Schuller I K 1999 *J. Magn. Magn. Mater.* **192** 203–32

[3] Kortright J, Awschalom D, Stöhr J, Bader S, Idzerda Y, Parkin S, Schuller I and Siegmann H C 1999 *J. Magn. Magn. Mater.* **207** 7–44

[4] Baran E J 1990 *Catalysis Today* **8** 133–51

[5] Haeni J H *et al.* 2004 *Nature* **430** 758–61

[6] White R L 1969 *J. Appl. Phys.* **40** 1061–9

[7] Geller S and Wood E A 1956 *Acta Cryst.* **9** 563–8

[8] Glazer A M 1975 *Acta Crystallogr.* A **31** 756–62

[9] Eibschütz M, Shtrikman S and Treves D 1967 *Phys. Rev.* **156**(2) 562–77

[10] Lyubutin I S, Dmitrieva T V and Stepin A S 1999 *J. Exp. Theor. Phys.* **88** 590–7

[11] Seo J W, Fullerton E E, Nolting F, Scholl A, Fompeyrine J and Locquet J P 2008 *J. Phys. Condens. Matter* **20**

[12] Hallsteinsen I, Boschker J, Nord M, Lee S, Rzchowski M, Vullum P, Grepstad J, Holmestad R, Eom C and Tybell T 2013 *J. Appl. Phys.* **113**

Analysis of interfacial structure and chemistry in FeV_2O_4-based heterostructures on (001)-oriented $SrTiO_3$

Y Zhou[1], D Zhou[1], D R G Mitchell[2], N Valanoor[1] and P Munroe[1]
[1] School of Materials Science and Engineering, University of New South Wales, Sydney NSW 2052, Australia
[2] Electron Microscopy Centre, Australian Institute for Innovative Materials, University of Wollongong, North Wollongong NSW 2522, Australia

E-mail: yanyu.zhou@student.unsw.edu.au

Abstract. In this study, we investigate the interfacial structures and chemistry of FeV_2O_4 (FVO)/ $SrTiO_3$ (STO) and $FeV_2O_4/La_{0.33}Sr_{0.67}MnO_3/SrTiO_3$ (FVO/LSMO/STO) heterostructures, in which FVO is epitaxially grown on both a (001)-oriented $SrTiO_3$ substrate and a LSMO-buffered substrate by pulsed laser deposition (PLD). By combining data from transmission electron microscopy (TEM) imaging together with high-resolution scanning transmission electron microscopy (STEM) imaging, the interfaces between FVO and STO, and FVO and LSMO are found to be coherent and semi-coherent, respectively, in spite of more than 8% of strain at the interface. Energy-dispersive X-ray spectroscopy (EDS) analysis suggests that localised diffusion occurs at both the FVO/STO and the FVO/LSMO interfaces.

Introduction

Non-collinear multiferroics (NCMs) are a new class of functional oxides with significant potential in spintronic and nanoelectric application devices, due to their magnetoelectric coupling effects. NCMs produce magnetoelectric coupling by possessing both ferroelectric and ferromagnetic ordering at low temperatures. An NCM, such as FeV_2O_4 (FVO), is a multiferroic which undergoes successive phase transitions and possesses both ferroelectric and ferrimagnetic moments at low temperatures [1, 2].

FVO, in thin film form, has attracted considerable research attention in recent years [3]. Thin-film FVO-based heterostructures exhibit novel properties thought to be associated with the interactions between lattice strain, electronic spin and orbital bonding [2]. Generally, a NCM thin film grown on a ferromagnetic or piezoelectric substrate will lead to large electrical and magnetic properties of the constituent epitaxial thin films [4]. In such thin film heterostructures, the relative lattice parameters of both the substrate and buffer layers play an important role in determining the strain and the structure of the epitaxial thin films. Semi-coherent and incoherent interfaces lead to large dislocation densities, which degrade the ferroelectric properties of these thin films due to the generation of local strain fields [5]. Hence, our aim is to explore the interfacial morphology and chemistry of the FVO-based heterostructures by transmission electron microscopy (TEM) techniques. In this report, epitaxial FVO-based heterostructures were produced by depositing the FVO films onto the STO substrate with, and without, a buffer layer of $La_{0.33}Sr_{0.67}MnO_3$

(LSMO) using pulsed laser deposition (PLD). These FVO/STO and FVO/LSMO/STO heterostructures were characterised using aberration-corrected scanning transmission electron microscope (STEM). Their interfacial structures were analysed by a combination of bright-field and high-angle annular dark-field (HAADF) imaging. The chemical interactions between the FVO and STO interfaces were investigated using energy-dispersive X-ray spectroscopy (EDS).

Experimental Methods

Two FVO/STO and FVO/LSMO/STO heterostructures were fabricated by PLD. Their crystal structures and surface morphologies were then examined by X-ray diffraction (XRD) and atomic force microscopy (AFM) prior to TEM sample preparation. Cross-sectional TEM specimens that were suitable for both low-magnification TEM and high-resolution STEM imaging were prepared by conventional focused ion beam (FIB) and tripod polishing/ion polishing methods. General microstructural features such as FVO film thickness of the two heterostructures were obtained by analysing FIB specimens using a Philips CM200 field emission TEM. Interfacial structures and chemical interactions of the two heterostructures were investigated by STEM bright-field and HAADF imaging and EDS using a JEOL JEM-ARM200F with tripod-prepared specimens.

Results & Discussion

XRD spectra from FVO/STO and FVO/LSMO/STO confirm the presence of FVO lattice planes in the two grown single-crystalline heterostructures. AFM scans show the mean roughness of the FVO films in the FVO/STO and FVO/LSMO/STO heterostructures was 1.3 nm and 1.5 nm respectively. For brevity neither the XRD nor AFM data are shown here.

A STEM cross-sectional bright-field image of a FVO/STO specimen oriented down at the [100]-zone axis is shown in Figure 1(a). The FVO film is observed to be uniformly distributed along the interface and approximately 75 nm in thickness with a mean surface roughness consistent with the AFM scan. Strain contrast is visible both at the interface and within the FVO film due to the more than 8% of lattice mismatch (a_{FVO} = 0.846 nm [6]; a_{STO} = 0.3905 nm [7, 8]). Higher resolution bright-field and HAADF images at the FVO/STO interface are shown in Figure 1(b) and Figure 1(c) respectively. A reduced fast-Fourier transform (FFT) (Figure 1(b) inset) of a selected area encompassing both the FVO and STO regions, indicates the epitaxial nature and the relaxation of the FVO film in which, the in-plane of the FVO was parallel to that of STO (e.g. $(002)_{STO}//(004)_{FVO}$) and the out-of-plane FVO and STO were parallel to each other (e.g. $(020)_{STO}//(040)_{FVO}$).

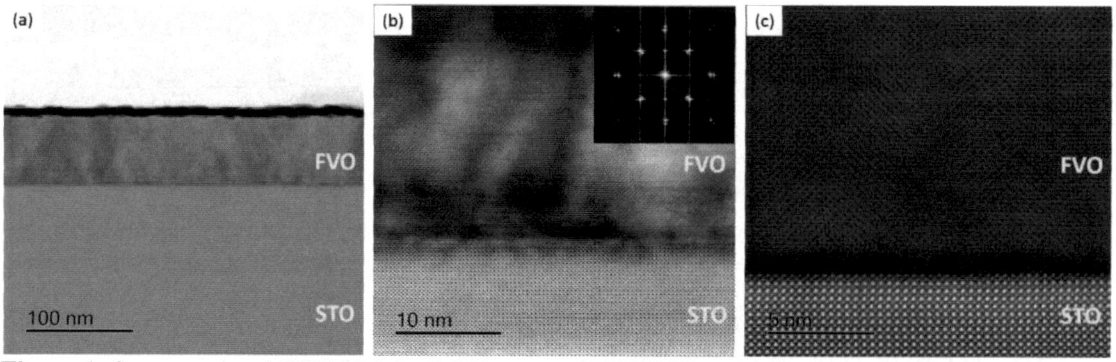

Figure 1. Cross-sectional STEM bright-field images of FVO/STO at (a) low magnification; and (b) high-magnification with reduce FFT inset at zone axis. (c) STEM HAADF image of FVO/STO.

Although no dislocations were observed, they are generally known for relieving the strain induced from the lattice mismatch between the film and the substrate. In order to investigate the possible presence of misfit dislocations, bright-field images of FVO/STO specimen were captured under two beam diffraction conditions, as displayed in Figure 2(a) and 2(b). In Figure 2(a), Moiré fringes were observed along the FVO/STO interface. The average spacing of the Moiré fringes is measured in Figure 2(b) to be 2.98 nm. It suggests the overlapping of the lattice planes

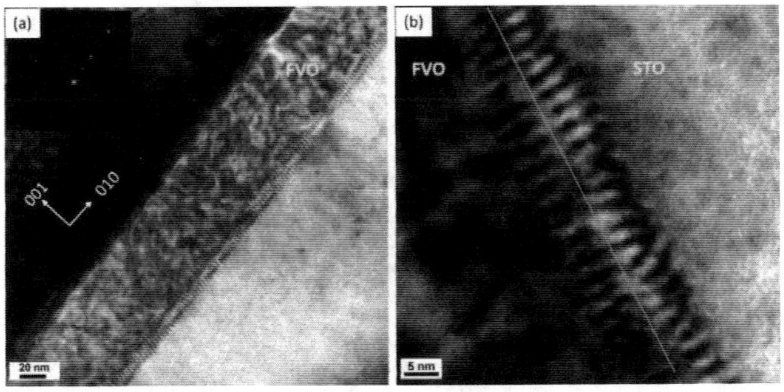

Figure 2. (a) Low-magnification bright-field image of FVO/STO under two beam diffraction conditions (g = 010). (b) High-magnification bright-field image of FVO/STO containing Moiré fringes with local spacing measurement inset.

Interfacial diffusion may also play an important role in retaining the epitaxial nature of the FVO film. An EDS line scan across the FVO/STO interface (Figure 3(a) and 3(b)) suggests localised atomic interdiffusion may occur in this region. As the linescan traverses from STO to FVO, the Fe-K and V-K intensities gradually increase over a distance of 3~4 nm, while the Ti-K and Sr-K intensities decrease over the same distance. This EDS line analysis was performed using a JEOL JEM-ARM200F in which the aberration-corrected TEM was operated at a voltage of 200 kV, with a point resolution that could be achieved to less than 0.08 nm. This suggests the gradual change of intensities of Fe-K and V-K is not a result of beam broadening, but rather a case of interdiffusion where Fe and V ions diffuse into the substrate lattice.

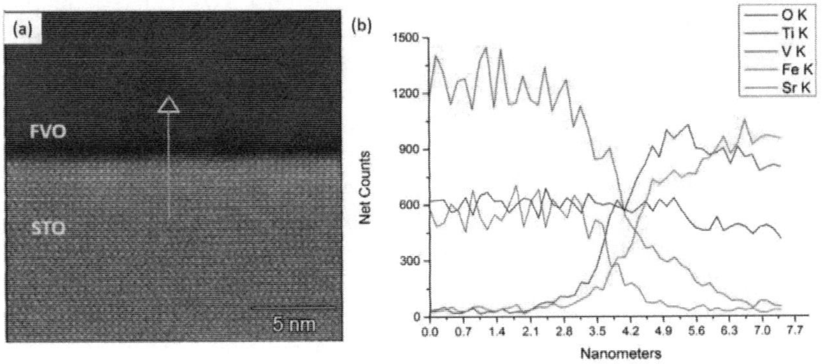

Figure 3. (a) EDS line analysis and scan direction at the interface of FVO/STO specimen. (b) EDS line scan results show gradual increase of the Fe-K and V-K intensities over a distance of 4 nm across the FVO/STO interface.

For the FVO/LSMO/STO sample, a series of STEM bright-field images were captured with the sample oriented parallel to the [100]-zone axis and shown in Figure 4. In Figure 4(a), although the FVO film uniformly grows onto of the LSMO-buffered substrate with a mean surface roughness of 1.5 nm, it appears to obtain low-angle grain boundaries. These low-angle boundaries are more evident at higher magnification (Figure 4(c)) where fringes are observed between neighbouring grains. The inset reduced FFT also shows the tilting of the FVO planes. Due to more than 8% lattice mismatch (a_{FVO} = 0.8460 nm [6]; a_{LSMO} = 0.3894 nm [9]), mismatch dislocations are introduced in the upper region of the LSMO as seen in Figure 4(b). An EDS line scan was also carried out across the FVO/LSMO interface (not shown here) to investigate the changes of intensities of Fe and V across this region. Again, the intensities of Fe-K and V-K signals drop gradually over a distance of 3.5 nm while the Mn-K signal increases, which suggest the occurrence of localised diffusion at this particular interface.

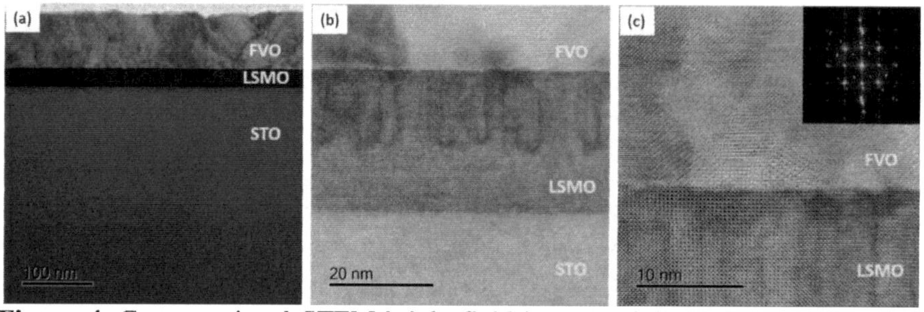

Figure 4. Cross-sectional STEM bright-field images of FVO/LSMO/STO at (a) low magnification and (b) high-magnification showing mismatch dislocations in the LSMO upper layer. (c) STEM bright-field image of FVO/LSMO interface with reduce FFT inset.

Conclusion

Cross-sectional studies of epitaxial FVO/STO and FVO/LSMO/STO by TEM and STEM imaging reveal interdiffusion at interfaces and misfit dislocations play parts to relieve strain due to large lattice mismatches. However, the exact nature of diffusion needs further investigation. Further, EELS studies on the interfaces is required.

Acknowledgement

This research has been supported by the Australia-India Strategic Research Fund (AISRF). The authors would like to acknowledge the facilities, and the scientific and technical assistance, of the Australian Microscopy & Microanalysis Research Facility at the Electron Microscopy Unit, University of New South Wales. This experiment also used equipment funded by the Australian Research Council (ARC) – Linkage, Infrastructure, Equipment and Facilities (LIEF) located at the University of Wollongong Electron Microscopy Centre.

References

[1] Nii Y, Sagayama H, Arima T, Aoyagi S, Sakai R, Maki S, Nishibori E, Sawa H, Sugimoto K, Ohsumi H and Takata M 2012 *Phys. Rev. B* **86** 125142
[2] Liu N, Zhao K H, Shi X L and Zhang L W 2012 *J. Appl. Phys.* **111** 124112-5
[3] MacDougall G J, Garlea V O, Aczel A A, Zhou H D and Nagler S E 2012 *Phys. Rev.* B **86** 060414(R)
[4] Zheng R K, Wang Y, Liu Y K, Gao G Y, Fei L F, Jiang Y, Chan H L W, Li X M, Luo H S and Li X G 2012 *Mater. Chem. Phys.* **133** 42-46
[5] Marin L W and Ramesh R 2012 *Acta Mater.* **60** 2449-70
[6] Zhang Q, Ramazanoglu M, Chi S, Liu Y, Lograsso T A and Vaknin D 2014 *Phys. Rev. B* **89** 224416
[7] Schlom D G, Chen L -Q, Eom C -B, Rabe K M, Streiffer S K and Triscone J -M 2007 *Annu. Rev. Mater. Res.* **37** 589-626
[8] Martin L W, Chu Y -H and Ramesh R 2010 *Mater. Sci. Eng. R: Rep.* **68** 89-133
[9] Lyonnet R, Maurice J L, Hÿtch M J, Michel D and Contour J P 2000 *Appl. Surf. Sci.* **162-163** 245-9

New hexagonal perovskite with Mn^{4+} and Mn^{5+} at distinct structural positions

N V Tarakina[1,2], A P Tyutyunnik[3], G V Bazuev[3], A D Vasiliev[4,5], I F Berger[3], C Gould[2], I V Nikolaenko[3]

[1]The NanoVision Centre, School of Engineering and Materials Science, Queen Mary University of London, Mile End, London E1 4NS, United Kingdom

[2]Experimentelle Physik III, Physikalisches Institut and Wilhelm Conrad Röntgen - Research Centre for Complex Material Systems, Universität Würzburg, Am Hubland, D-97074 Würzburg, Germany

[3]Institute of Solid State Chemistry, Ural Branch of the Russian Academy of Sciences, 91 Pervomayskaya, 620990, Ekaterinburg, Russia

[4]L.V. Kirensky Institute of Physics, Siberian Branch of Russian Academy of Sciences, 50 Akademgorodok, 660036, Krasnoyarsk, Russia

[5]Siberian Federal University, Svobodny pr. 79, 660041 Krasnoyarsk, Russia

E-mail: n.tarakina@qmul.ac.uk

Abstract. A new hexagonal perovskite, $Ba_7Li_{1.75}Mn_{3.5}O_{15.75}$, has been synthesised using microwave-assisted solid-state synthesis. Its crystal structure has elements typical for the layered hexagonal perovskites and quasi-one-dimensional oxides, hence representing a new polytype. Structural solution based on simultaneous refinement of X-ray and neutron diffraction data shows that $Ba_7Li_{1.75}Mn_{3.5}O_{15.75}$ crystallizes in a hexagonal unit cell with parameters a = 5.66274(2) Å and c = 16.7467(1) Å (V = 465.063(4) Å^3). Columns of face-shared octahedra occupied by Mn^{4+}, Li^+ cations and vacancies along the c axis are separated in the ab plane by barium atoms, so that every sixth layer, the coordination of Mn^{5+} and Li^+ changes to tetrahedral. Separation of Mn^{4+} and Mn^{5+} cations in two distinct structural positions makes the structure unique. A scanning transmission electron microscopy study revealed the formation of a rhombohedrally centered supercell, which might be attributed to the ordering of manganese and lithium atoms among cationic sites.

1. Introduction

$AMnO_3$ (A = alkali-earth element) perovskites display a variety of interesting magnetic and electronic phenomena (metallic ferromagnetism, charge or orbital ordering, etc.), which result from competing charge, exchange and phonon interactions [1]. A fine-tuning of these competing interactions and thus control of the functional properties of these materials can be realized by substituting either A-site cations or Mn with other cations (of different sizes/charges). Substitution changes the distribution of the charges (ratio between Mn cations in different oxidation states) or leads to the formation of different polymorphs (changing the ratio of corner-sharing (cubic) to face-sharing (hexagonal) perovskite layers within one structure): 6H-$Ba_{0.7}Sr_{0.3}Ru_{1-x}Mn_xO_3$ ($0.2 \le x \le 0.4$) [2], 8H-$Ba_8Ta_6NiO_{24}$ [3],

10H-$Ba_5Sn_{1.1}Mn_{3.9}O_{15}$ [4], 12H-$(BaSr)_{0.5}Mn_4Cr_2O_{17}$ [5], 9R-$BaRu_{1-x}Mn_xO_3$ ($0 \leq x \leq 0.90$) [6], 12R-$Ba_3NdMn_2O_9$ [7], 16R-$Ba_4Ca_{1-x}Mn_{1+x}O_{12-\delta}$ [8], etc. — see Figure. 1.

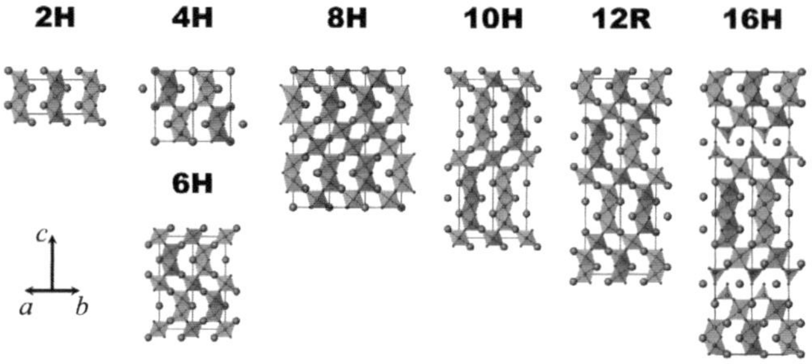

Figure 1. Crystal structures of some polytypes of $BaMnO_3$ and different polymorphs obtained by substitution of Mn by other elements (A'). Red and blue spheres indicate Ba and oxygen sites, respectively. Positions indicated by green polyhedra are occupied by manganese and A' atoms.

Substitution of the Mn^{4+} ions in $BaMnO_3$ by cations of elements with smaller valence leads to the formation of compounds with mixed or enhanced average degree of oxidation of manganese such as the new family of layered structures of the general formula $Ba_{5+n}Ca_2Mn_{3+n}O_{3n+14}$, the hexagonal 6-layer perovskite $Ba_3ErMn_2O_9$ [9], as well as the quasi-one-dimensional compounds $Ba_4NaMn_2O_9$ [10] and $Sr_4LiMn_2O_9$ [11]. A structure with distinct atomic positions for the manganese Mn^{4+} and Mn^{5+} ions has been found for $Ba_7Ca_2Mn_5O_{20}$ but not for the aforementioned quasi-one-dimensional oxides. In this work, we focused on synthesis, crystal structure and magnetic properties of the new $Ba_7Li_{1.75}Mn_{3.5}O_{15.75}$ oxide.

2. Results and Discussion

$Ba_7Li_{1.75}Mn_{3.5}O_{15.75}$ was been synthesized for the first time, by means of a microwave-assisted solid-state method. A stoichiometric mixture of the initial reagents $BaCO_3$ (99.9%), Li_2CO_3 (99.99%) and MnO_2 (99.9%) was calcined at 950°C for 1 hour in a microwave furnace (generator power of 700 W and frequency of 2450 MHz), then ground, pressed into a pellet and annealed at 1030°C for 2 hours with intermediate grinding after 1 hour of annealing.

Table 1. Atomic coordinates and isotropic thermal parameters ($U_{iso} \times 100$, Å^2) for $Ba_7Li_{1.75}Mn_{3.5}O_{15.75}$.

Atom	Site§	x/a	y/b	z/c	Fraction	$U_{iso} \times 100$
Ba(1)	1a	0.0	0.0	0.0	1.0	3.6(1)
Ba(2)	2d	1/3	2/3	0.5724(2)	1.0	2.63(9)
Ba(3)	2d	2/3	1/3	0.6964(2)	1.0	2.3(1)
Ba(4)	2d	1/3	2/3	0.8263(1)	1.0	1.9(1)
Mn/Li(1)	2c	0.0	0.0	0.3509(4)	0.79/0.21(1)	3.9(2)*
Mn/Li(2)	1b	0.0	0.0	1/2	0.67/0.33(1)	3.9(2)*
Mn/Li(3)	2d	2/3	1/3	0.9085(6)	0.80/0.20(1)	3.9(2)*
Mn/Li(4)	2c	0.0	0.0	0.194(2)	0.005/0.995(8)	3.9(2)*
O(1)	6i	0.1785(8)	0.8215(8)	0.1307(4)	0.88(1)	1.73(8)*
O(2)	6i	0.8443(8)	0.689(2)	0.5677(5)	0.718(8)	1.73(8)*
O(3)	6i	0.1511(7)	0.302(1)	0.7083(4)	1.0	1.73(8)*
O(4)	2d	2/3	1/3	0.004(1)	0.59(2)	1.73(8)*

§Symmetry multiplicity and Wyckoff symbol; *Thermal parameters of manganese/lithium and oxygen atoms were constrained as single variables.

X-ray and neutron powder diffraction patterns were indexed in a hexagonal unit cell with parameters a = 5.66274(2) Å and c = 16.7467(1) Å. No systematic absences were observed, so all primitive trigonal space groups can be chosen. A crystal structure similar to the structure of the quasi-1D $Sr_4LiMn_2O_9$ oxide, but with positions of one alkali-earth atom and one Mn atom, exchanged (Figure 2, a,b) was used as a starting model. In the final model, columns of face-sharing octahedra occupied by Mn^{4+} and Li^+ oriented along the c axis are cut in short chains of five by Ba atoms, while in the barium columns every sixth layer is now occupied by Mn^{5+}/Li^+ in tetrahedral coordination. For the resulting model the *P*-3*m*1 (164) space group was chosen. To get more accurate positions of the lithium and the oxygen atoms, the neutron (ND) and X-ray diffraction (XRD) data were refined simultaneously (Table 1). The presence of the structural elements described above was confirmed by high-resolution transmission electron microscopy; however, selected-area electron diffraction (SAED) patterns showed the formation of a rhombohedrally centered supercell with unit cell parameters a = 10 Å, c = 51.3 Å. Extinction conditions point to five possible space groups: *R*3 (146), *R*-3 (148), *R*32 (155), *R*3*m* (160), *R*-3*m* (166), Figure 2, e-g. An analysis of the intensity distributions on high-angle annular dark field scanning transmission electron microscopy (HAADF-STEM) images suggests that the formation of a supercell is related to additional ordering of Mn and Li atoms in the *ab* plane (Figure 2, c,d). In particular, according to the structural model obtained from XRD all atomic columns along the $[1\text{-}10]_{P\text{-}3m1}$ ($[100]_{R\text{-}3m}$) direction should give equal contrast on the HAADF-STEM images, however on the recorded images two columns of higher intensity alternate with one column of lower intensity, Figure 2 (h, i). Simulations of HAADF-STEM images did not provide an unambiguous solution of the cation distribution; additional annular bright-field imaging or atomic-column resolved Li-K edge spectrum imaging are required to obtain a full description of the supercell. Here and in the following we will refer to such a set of three columns as one block. The described cation ordering explains the increase of the a parameter of the unit cell by a factor of $\sqrt{3}$ and a rotation of the crystal axes by 30° with respect to the primitive trigonal lattice.

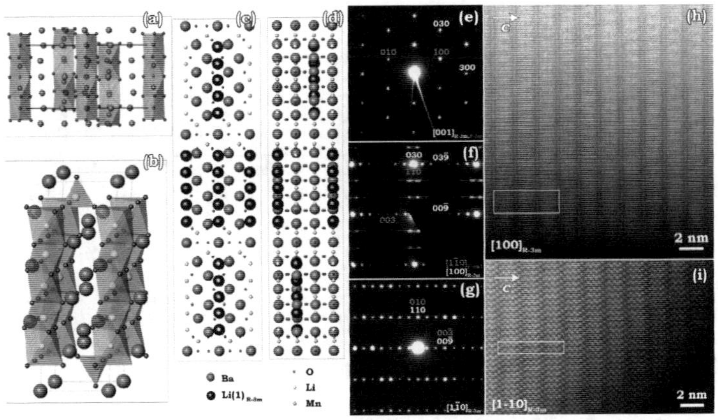

Figure 2. Schematic drawings of quasi-one-dimensional $Ba_4NaMn_2O_9$ oxide (a), the average structure of $Ba_7Li_{1.75}Mn_{3.5}O_{15.75}$ obtained from XRD (b), the rhombohedral superstructure of $Ba_7Li_{1.75}Mn_{3.5}O_{15.75}$ in projections on $(110)_{R\text{-}3m}$ (c), $(1\text{-}10)_{R\text{-}3m}$ (d). SAED patterns taken along the main zone axis (e-g). HAADF-STEM images taken along $[100]_{R\text{-}3m}$ (h) and $[1\text{-}10]_{R\text{-}3m}$ (i) directions.

A periodic stacking of blocks along the c direction with shifts equal to 2/3 of the width of the block would result in a perfect rhombohedral supercell. However, occasional shifts over 1/3 (= -2/3) have been observed in the HAADF-STEM images, destroying the perfect rhombohedral symmetry of the crystal and causing the presence of diffuse streaks in the SAED patterns along the $[100]_{R\text{-}3m}$ direction as well as problems with refining XRD and ND data using the supercell model. The defect structure can be described either through the formation of twin planes perpendicular to the c direction or through stacking faults along the c direction.

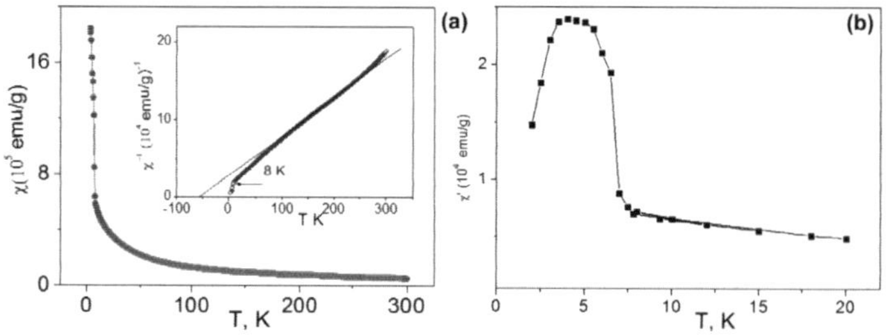

Figure 3. Temperature dependence of the DC magnetic susceptibility χ (a) and of the real component of the AC magnetic susceptibility χ' for $Ba_7Li_{1.75}Mn_{3.5}O_{15.75}$. The inset shows the χ^{-1}(T) dependence.

Static magnetic susceptibility measurements were carried out at H = 0.3 kOe in the temperature range 3-300 K (Figure 3). Between 100 and 220 K the magnetic susceptibility χ obeys the Curie-Weiss law χ = C/(T-θ), where C is the Curie constant (2.40 cm^3 K/mol), and θ is the Weiss temperature (θ = -55 K). Above 220 K and below 100 K, deviations from a straight-line behavior are observed, pointing to an antiferromagnetic exchange interaction between the paramagnetic manganese ions in $Ba_7Li_{1.75}Mn_{3.5}O_{15.75}$, similar to those found in the $Sr_4AMn_2O_9$ series (A = Li, Zn, Ni, Cu) [11-13]. For the determination of the reason of the sharp χ increase below 8 K, AC magnetic susceptibility measurements were performed. The real χ' and imaginary χ'' components of the magnetic susceptibility were determined for alternating-magnetic-field amplitudes of up to 4 Oe at a frequency of f = 80 Hz and temperatures in the range T = 2 - 20 K (Fig. 3). Below 8 K, χ' increases (just like χ), until it reaches a maximum at 5 K, below which it decreases. The occurrence of a maximum in the dependence of χ' on T is a signature of a magnetic transition, the nature of which is possibly related to antiferromagnetic properties below the transition temperature.

3. Conclusions

$Ba_7Li_{1.75}Mn_{3.5}O_{15.75}$ represents a new hexagonal perovskite polytype with structural elements typical for layered hexagonal perovskites and the quasi-one-dimensional oxides, and with Mn^{4+} and Mn^{5+} cations distinctly distributed between structural positions. The temperature dependence of the magnetic susceptibility reveals the presence of antiferromagnetic exchange interactions between the paramagnetic manganese ions in the hexagonal columns and suggests antiferromagnetic order < 5 K.

4. Acknowledgments

X-ray powder diffraction studies were carried out at the Center for collective use "X-ray structure analysis" at the Institute of Solid State Chemistry, UrB RAS (Ekaterinburg, Russia).

References

[1] Tokura Y and Nagaosa N 2000 *Science* **288** 462
[2] Yin C, Li G, Lin J and Attfield J P 2009 *Chem. Asian J.* **4** 969
[3] Shpanchenko R V, Nistor L, Van Tendeloo G *et al* 1995 *J. Solid State Chem.* **114** 560
[4] Yin C, Li G, Jin T, Tao J, Richardson J, Loong C-K *et al* 2010 *J. Alloys Compounds* **489** 152
[5] Clark J H and Hayward M A 2008 *Chem. Mater.* **20** 4612
[6] Yin C, Li G, Kockelmann W A, Lin J and Attfield J P 2009 *Phys. Rev. B* **80** 094420
[7] Yang H, Tang Y K, Yao L D, Zhang W, Li Q A *et al* 2007 *J. Alloys Compounds* **432** 283
[8] Floros N, Michel C, Hervieu M and Raveau B 2000 *Chem. Mater.* **12** 3197
[9] Rabbow C and Müller-Buschbaum H Z 1994 *Naturforsch.* **498** 1277
[10] Quarez E, Roussel P L, Pérez O, Leligny H *et al* 2004 *Solid State Sci.* **6** 631
[11] Bazuev G V, Tyutyunnik A P, Berger I F *et al* 2011 *J. Alloys Compounds* **509** 6158
[12] Bazuev G V, Krasilnikov N V and Kellerman D G 2003 *J. Alloys Compounds* **352**, 190
[13] Abed A El, Gaudin E, Darriet J and Whangbo M-H 2002 *J. Solid State Chem.* **163**, 513

Electron Energy Loss Spectroscopy of a Chiral Plasmonic Structure

G W Paterson[1], A Karimullah[2], S D R Williamson[1], M Kadodwala[2] and D A MacLaren[1]

[1] SUPA, School of Physics and Astronomy, University of Glasgow, Glasgow G12 8QQ, UK.
[2] School of Chemistry, University of Glasgow, Glasgow G12 8QQ, UK.

E-mail: gary.paterson@glasgow.ac.uk

Abstract. A detailed analysis of the plasmonic excitations within a nanopatterned gold chiral biosensor element, measured by scanning transmission electron microscopy electron energy loss spectroscopy, is presented. We discuss aspects of data acquisition, processing, analysis and simulation. The localised surface plasmonic resonance modes in the structure are extracted using non-negative matrix factorisation and we use simulations to correlate notable deviations from the idealised spectrum to nanometric fabrication imperfections. The methodology presented has wide applicability to a variety of metamaterials.

1. Introduction

Plasmonic metamaterials have a diversity of applications, spanning optoelectronics, sensors and biomedicine. Their functionality derives from a precise control of patterned media down to the nanoscale, which can be difficult to assess with optical techniques. Energy loss spectroscopy (EELS) performed in a scanning transmission electron microscope (STEM), on the other hand, has ample spatial resolution but one can struggle to assess the plasmonic energy range associated with optical excitations because of the energy spread of the source and the proximity of the plasmon signal to the zero loss peak (ZLP). Motivated by recent work showing that chiral metamaterials can be used as a sensor of biomolecules [1], here we study a chiral nanostructure with STEM-EELS. A discussion of the chiral fields produced by such structures can be found elsewhere [2]. We present direct, nanometric maps of the localised surface plasmon resonance (LSPR) modes. Comparisons of experimental results to simulations enable the correlation of fabrication imperfections to the plasmonic properties.

2. Methodology

The structure under study, an approximately 60 nm thick, 500 nm diameter gold 'nanoswirl' was formed on a silicon nitride (SiN) window using electron beam lithography and e-beam evaporation with a 2 nm Ti adhesion layer. Figures 1(a) and (b) show the typical structure of the Au plasmonic element, which is markedly different to the ideal structure shown in Fig. 1(e). The edge-profile of the fabricated structure is tapered due to the metal deposition method used and the central region is enlarged because of overexposure of the resist during lithographic patterning; both aspects are a consequence of the relatively small feature sizes in the structure. In the central region, the thickness is relatively uniform and close to the nominal value, but notable thickness variations and a loss of edge definition occur along the arms, partly due to the Au grain structure. Deviations from the ideal square-edged profile of the geometric design break the symmetry of the nanostructure and alter its plasmonic response, so it is essential to assess their impact. To do this we performed EELS measurements in a probe corrected JEOL ARM200F scanning transmission electron microscope operated at 200 keV and equipped with a cold field emission electron gun and a GIF Gatan Quantum ER spectrometer. The EELS data were collected at a dispersion of 0.025 eV/channel and spatial resolution of 3 nm. The dwell time was 50 ms and the convergence and collection semi-angles were 29 and 36 mrad. The thickness of the structure was extracted using the t/λ method [3], in which t is the thickness and λ is the inelastic mean free path, and was then used to form a mesh for use in the MNPBEM Matlab toolbox with EELS classes [4]. All data were analysed and processed in Python using the Hyperspy package [5].

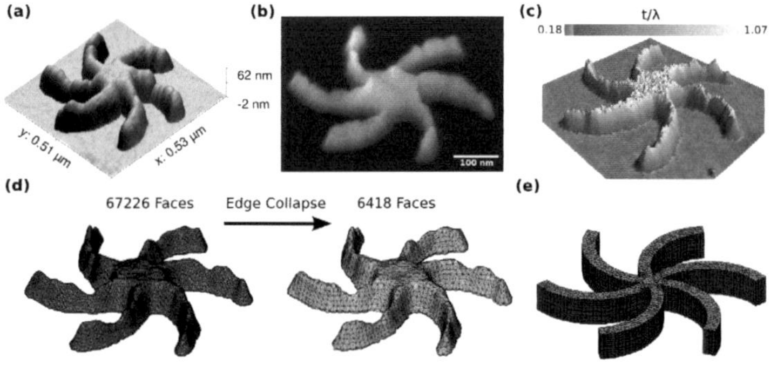

Figure 1. Typical nanoswirl plasmonic elements imaged by (a) AFM, (b) SEM (17.5 kV, 52° tilt), and, (c) EELS t/λ. In (d) and (e) are surface meshes of the t/λ data for the experimental structure and the surface mesh of the ideal 60 nm thick nanoswirl. Data were taken of different elements of the design.

3. Results & Discussion

An unexpected artefact in the data was a small lateral shift of the STEM diffraction disk as the beam scanned across the sample, because the projection system was not perfectly coupled to the microscope's diffraction plane. This can usually be corrected by post acquisition realignment of the EELS spectra, shifting spectra a few pixels along the dispersion direction. However, the weak plasmon signal lies close to the intense ZLP, so an artefact can arise if it overlaps with the ZLP afterglow of a previous spectrum. The shift in the beam position is approximately described by a plane, as shown in Figure 2(a). Thus, during a line-scan with small step size, continual drift of the beam results in the afterglow lying within the ZLP and the effect is only to broaden the peak slightly. However, at the beginning of a new line-scan, a larger offset in the beam position occurs. In the example data of Figure 2(a), the maximum shift of the ZLP is around 3 eV and results in a peak in subsequent spectra at similar energies and of similar intensities to the plasmon signal, as shown in Figure 2(b). To counteract this, spatial drift correction was employed at the end of every scan line, which allowed sufficient time for the afterglow to decay to insignificant levels. The effect of the artefact is seen as horizontal banded lines of decaying intensity in Figure 2(c), for which drift correction was used every fourth row. Only those rows acquired immediately after drift correction are free from afterglow artefacts. An alternative method for reducing the impact of the artefact would be to choose (in our case to reverse) the scan direction so that the afterglow falls on the energy gain side of the ZLP, which also has the advantage of not decreasing the rate at which data can be collected.

Next we discuss how the EELS data were processed to extract the plasmonic properties. The energy resolution of the EELS measurements is determined by the full width half maximum (FWHM) of the ZLP and is typically around 0.35 eV. To improve the energy resolution, Richardson-Lucy deconvolution [6] was used to narrow the FWHM of the ZLP to 0.11 eV. The deconvolved

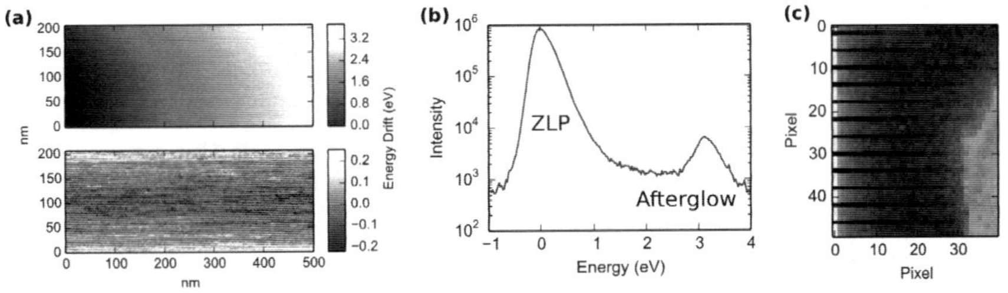

Figure 2. (a) Example sections of ZLP energy drift with scan position as measured (top) and after plane subtraction (bottom). (b) Spectra showing an afterglow artefact at the start of a new line-scan. (c) Maximum intensity on a log scale in a window covering the energy range of the artefact at different scan positions. The horizontal bands in (c) are due to drift correction allowing time for the afterglow to deminish to insignificant levels.

data were then processed using non-negative matrix factorisation [7] (NMF), a multivariate decomposition technique which has been successfully used to extract plasmonic modes [8]. An advantage that NMF holds over other decomposition protocols is that it restricts the components to be positive and so is constrained to produce spectral components that are often more representative of the real excitations. Typical results of performing factorisation of the nanoswirl dataset with 4 components is shown in Figure 3, with the decomposition factors (DFs) and loadings (DLs) in panels (a) and (b). The decomposition components are readily identified from the spatial distributions [Fig. 3(b)] and compare reasonably well with excitations observed in the unprocessed data. The first component, DL0, is a background signal localised within the gold and corresponds to a broadening of the ZLP. The fourth component is delocalised across the entire structure and corresponds to the leading edge of a bulk plasmon. Of greater significance are DL1 and DL2, modes with losses concentrated at the tips and base of the swirl, respectively. These can be seen as diffuse intensity outside of the main structure, at locations indicated by red arrows. The double peak in component 2 indicates that two modes are being extracted as one component. An example of the influence of the structural imperfections is the greater loss at the tips of the vertical arms in DL1, which would not occur in a symmetric structure.

To explore the influence of the structural imperfections on the plasmonic modes, we performed simulations of different structures using the MNPBEM toolbox [4]. The mesh of the ideal structure was generated directly in the toolbox and is shown in Figure 1(e). Forming a surface mesh of the exact structure requires additional tools and we did this by first determining the structure from the sample thickness, measured using the t/λ method [3]. Figure 1(c) shows the t/λ values calculated after extrapolating the spectra to higher energies using a power law. With the substrate thickness removed by subtracting a plane fitted to the masked surface to select

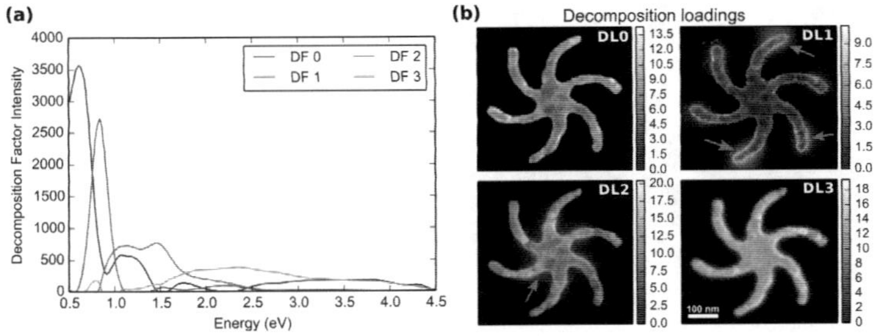

Figure 3. NMF of EELS spectra from a Au nanoswirl metamaterial element, showing (a) the factors and (b) the loadings corresponding to the characteristic spectra and spatial distributions, respectively.

only the bare SiN region, the Au t/λ values were then converted to absolute thicknesses using the Iakoubovskii form for $\lambda(\rho)$ [9] with a density, ρ, of 19.3 g/cm^3, giving a λ of 90.9 nm. To create the surface mesh, the metal thickness map was converted into a discretised binary volume after gently smoothing the data by 2-D gaussian convolution, and then to a 2-D surface mesh using a marching cubes algorithm. This mesh was then simplified by edge collapse, with the resulting meshes shown in Fig. 1(d).

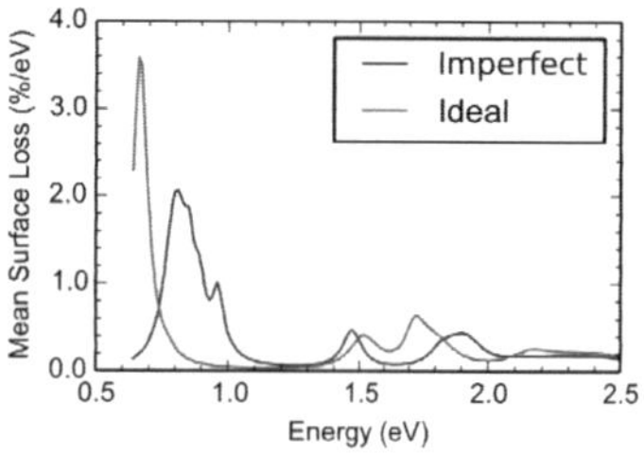

Figure 4. Simulated surface loss spectra from the imperfect and ideal structures, using the MNPBEM toolbox with tabulated values for Au permittivity. The substrate and adhesion layer were omitted for computational ease.

Finally, in Figure 4 the results of the simulations are compared. As a direct result of symmetry reduction due to structural imperfections, there are significant changes in the peak intensities and energies. The first peak is particularly strongly affected, with the single LSPR mode in the ideal structure split into a number of modes in the imperfect structure. The simulations of the imperfect structure compare favourably with the experimental data of Fig. 3. The different mode energies in the simulation may be related to the omission of the substrate and adhesion layer in the simulated structures.

4. Conclusions

The impact of structural imperfections on the plasmonic excitations of nanostructured materials can be assessed directly using EELS. Having knowledge of the precise, imperfect structure under study allows the effect of structural imperfections to be determined through simulation. Our results demonstrate that EELS is invaluable for simultaneously characterising the structural and the plasmonic properties, both in space and energy, of such nanostructures. With the advent of monochromated TEM instruments, the prospect for extending this work to a multitude of plasmonic metamaterial elements is an exciting one.

Acknowledgements

This work was funded by the Engineering and Physical Sciences Research Council (EPSRC) of the UK (EP/L001969/1). Original data files are available at: http://dx.doi.org/10.5525/gla.researchdata.195.

References

[1] Hendry E, Carpy T, Johnston J, Popland M, Mikhaylovskiy R V, Lapthorn A J, Kelly S M, Barron L D, Gadegaard N and Kadodwala M 2010 *Nat. Nano* **5** 783
[2] Schäferling M, Dregely D, Hentschel M and Giessen H, 2012 *Phys. Rev. X* **2** 031010
[3] Egerton R F 2011 *Springer, New York*
[4] Hohenester U 2014 *Computer Physics Communications* **185** 1177
[5] Hyperspy: A toolbox for hyperspectra data analysis *http://hyperspy.org*
[6] A. Gloter A, Douiri A, Tencé M and Colliex C 2003 *Ultramicroscopy* **96** 385
[7] Lee D D and Seung H S 1999 *Nature* **401** 788
[8] Nicoletti D, Peña F. de la, Leary R K, Holland D J, Ducati C and Midgley P A 2013 *Nature* **502** 80
[9] Iakoubovskii K, Mitsuishi K, Nakayama Y, Furuya K 2008 *Microsc. Res. Tech.* **71** 626

Self-consistent absorption correction for quantitative energy-dispersive X-ray spectroscopy of InGaN layers in analytical transmission electron microscopy

T Walther* and X Wang

Department of Electronic and Electrical Engineering, University of Sheffield, Mappin Building, Mappin Street, Sheffield S1 3JD, UK

*Email: t.walther@sheffield.ac.uk

Abstract. A new method of absorption correction for energy-dispersive X-ray spectroscopy in a transmission electron microscope is tested on InGaN samples. We simulate the effective *k*-factor for the In L line with respect to Ga L or Ga K and plot this as a function of the Ga K/L intensity ratio, which can be directly measured from experimental spectra. This basically performs an internal self-consistency check in the quantification using differently absorbed X-ray lines, which is in principle equivalent to an absorption correction as a function of specimen thickness but has the practical advantage that neither specimen thickness nor density or mass-thickness of the specimens need actually be measured.

1. Introduction

Energy-dispersive X-ray spectroscopy (EDXS) of thin foil cross-sections in a transmission electron microscope (TEM) is an important tool to determine local elemental concentrations and often relies on simple ratio measurements [1]. Quantitative measurements in transmission geometry necessitate corrections for absorption, fluorescence and detector efficiency of the individual X-ray lines where estimates of foil thickness and density are usually required for absorption correction [2]. The thickness of the specimen can be difficult to measure experimentally, and if one knew the density precisely then the chemistry could be directly inferred. The zeta-factor method [3] can determine concentrations and thickness iteratively but requires measurements of the electron dose and knowledge of the mass-absorption coefficients.

The method developed recently by us [4-7] does not need any of those external factors (thickness, density, mass-thickness) of the sample to be measured but relies on using a measured K/L intensity ratio as an internal absorption reference. In the following, we apply this to thin layers of $In_xGa_{1-x}N$.

2. Theoretical considerations

The original Cliff-Lorimer equation uses *k*-factors for weight% to relate concentrations, *c*, to measured intensities *I*, for two elements *i* and *j* in a very thin sample section in the form of

$$c_i/c_j = k_{ij}\ I_i/I_j \qquad \text{(equation 1).}$$

The *k*-factors in this equation are sensitivity factors that depend on the primary voltage, the X-ray fluorescence yield of the corresponding line and the energy-dependence of the sensitivity of the X-ray detector, compared to that of a standard. Only one X-ray line per element must be used to avoid double weightings. Most microscopists use the hardest X-ray line detectable for each element so absorption effects are kept minimal. For a multi-component alloy, $\sum c_i = 1$ and with simple relationships such as identity ($k_{ii}=1$), inversion ($k_{ij}= k_{ji}^{-1}$) and transferability ($k_{ij}=k_{i,Si} \times k_{Si,j}$) it can be shown that weight fractions are given by

$$c_i = \frac{I_i k_{i,Si}}{\sum_j I_j k_{j,Si}} \qquad \text{(equation 2)}$$

and atomic concentrations by

$$x_i = \frac{I_i k_{i,Si} / A_i}{\sum_j I_j k_{j,Si} / A_j} \qquad \text{(equation 3)}$$

where A_j denotes the atomic weight of element j. Considering a ternary semiconductor alloy such as $In_xGa_{1-x}N$ where the group-V sub-lattice is fixed we get for the group-III sub-lattice

$$x = \frac{I_{In} k_{In,Si} / A_{In}}{I_{In} k_{In,Si} / A_{In} + I_{Ga} k_{Ga,Si} / A_{Ga}} = \frac{I_{In} k_{In,Ga}}{I_{In} k_{In,Ga} + I_{Ga} A_{In} / A_{Ga}} \qquad \text{(equation 4)}$$

and from this

$$k_{In,Ga} = \frac{I_{Ga} A_{In}}{I_{In} A_{Ga} k_{In,Ga}} \times \frac{x}{1-x} \qquad \text{(equation 5)}$$

According to Cliff and Lorimer [1] this should only hold for a perfectly thin sample without any absorption; with absorption in a thicker sample exponential attenuation of every X-ray line occurs which is taken into account by including mass-thickness (or density), foil thickness and take-off angle as parameters. If instead we incorporate all these effects into k itself, we get effectively thickness-dependent k^*-factors

$$k^*_{InL,GaL} = x\, I_{GaL}\, A_{In} / [(1-x)\, I_{InL}\, A_{Ga}] \qquad \text{(equation 6a, for Ga L)}$$

and

$$k^*_{InL,GaK} = x\, I_{GaK}\, A_{In} / [(1-x)\, I_{InL}\, A_{Ga}] \qquad \text{(equation 6b, for Ga K)}$$

that are no longer constant factors but vary, as the intensities measured depend on absorption.

3. Simulations

For InGaN the X-ray lines of relevance are, listed in order of increasing energy in eV: N K (392); Ga Lα (1098) and Lβ (1125); In Lα (3287), $L\beta_1$ (3487), $L\beta_2$ (3714) and $L\gamma_1$ (3921); Ga Kα (9243) and Kβ (10264); In Kα (24139) and Kβ (27278). Figure 1 shows Monte Carlo simulations [8] of the dependence of $k^*_{InL,Ga}$ vs. Ga K/L ratio for $In_xGa_{1-x}N$ alloys of different indium concentrations. Curves for $x>0.5$ almost overlap, but for lower indium concentrations they differ, hence, no unique calibration curve as in the case of $In_xGa_{1-x}As$ [6] is obtained. However, all curves are monotonic as a function of Ga K/L ratio (all of them decreasing, due to absorption increasing with foil thickness), do not cross and yield higher $k^*_{InL,Ga}$ values for lower x values. This can be applied to determine the correct indium concentration of an alloy, x, iteratively: from the measured K/L ratio of a spectrum, any simulated calibration curve for x_{in} may be used as a starting point to estimate a value of k^*. Using this and the measured intensities, I, of the In L and Ga L (or In L and Ga K) lines as well as the atomic weights, A, a first estimate of x_{out} is obtained using equation (4). If $x_{out}>[<]\, x_{in}$, then the estimate of k^* was too big [small] and a calibration curve for a larger [lower] value of x_{in} must be used, which will yield a lower [higher] k^* value and thus a reduced [increased] x_{out}. Iteration will stop when $x_{out}\approx x_{in}$ at which point the correct indium concentration corresponds to the value of x_{out} for which the best fit simulation is available. The convergence is very quick: usually 2 or 3 iterations suffice. The approximate sign refers to the finite increments for which curves for x_{in} are simulated. In the case of figure 1 we used $\Delta x=0.2$ and already obtained sufficiently converged x values (with $\Delta x \leq 0.02$ from linear interpolation) after only 2 iterations. This procedure has several advantages over standard absorption corrections:

a) no estimates of foil thickness, density or mass-thickness are required, as the Ga K/L ratio can be calculated directly for each individual spectrum, serving as an internal standard [5];
b) results from quantification using the Ga L or the Ga K line are automatically self-consistent [6,7];
c) the detector sensitivity is no longer critical because the Ga K/L ratio as horizontal axis serves quasi as an internal self-correction, as shown in figure 2 where we simulated, for the example of x=0.5, the curves for an ideal detector with constant detection efficiency for all X-ray lines and our Si:Li detector with a thin window [7]. A reduced sensitivity for low energetic X-rays means, firstly, that the Ga K/L ratio will be increased, inducing a shift of all curves to the right. Secondly, as I_{GaL} / I_{InL} in equation 6a will be reduced, the monotonically decreasing curve for $k^*_{InL,GaL}$ will move down vertically so that in effect figure 2a will be almost unchanged. The ratio I_{GaK} / I_{InL} in equation 6b will correspondingly increase so that the monotonically increasing curve for $k^*_{InL,GaK}$ will move upwards vertically and, again, in effect it will be shifted only very little, by ~4% in figure 2b.

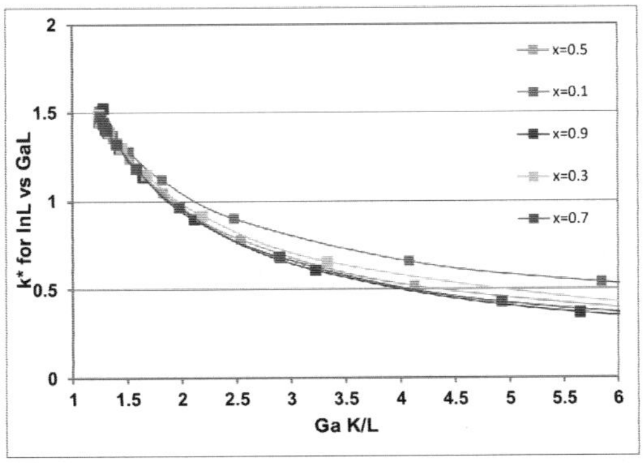

Figure 1a. CASINO simulations of $k^*_{InL,GaL}$ for $In_xGa_{1-x}N$, 10^6 electrons, 200kV, 25° take-off angle, ρ_{GaN} =6.15 g/cm^3, ρ_{InN} =6.91 g/cm^3, ideal detector, thicknesses up to~1500nm (x=0.1).

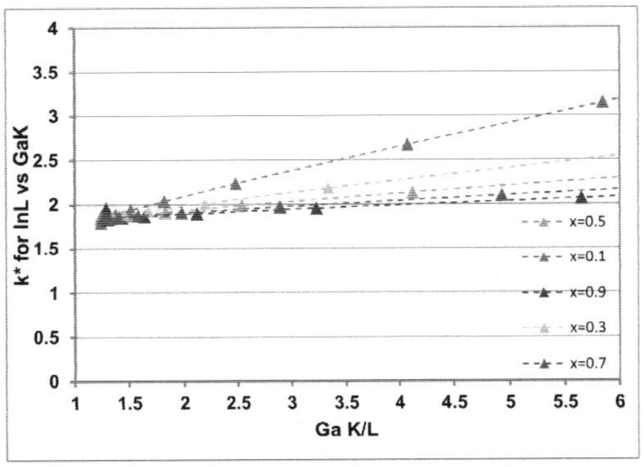

Figure 1b. CASINO simulations of $k^*_{InL,GaL}$ for various indium concentrations x, rest as for left.

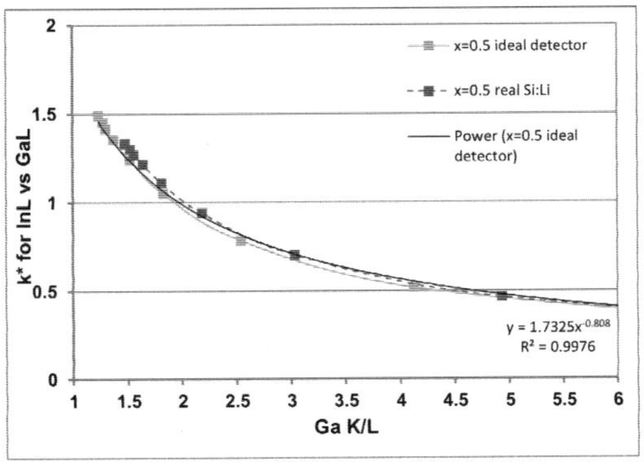

Figure 2a. CASINO simulations of $k^*_{InL,GaL}$ for $In_xGa_{1-x}N$ with ideal and Si:Li detector with modelled ratios of detector efficiencies of $\varepsilon_{GaK}/\varepsilon_{GaL}$=1.196 and $\varepsilon_{GaL}/\varepsilon_{InL}$=0.896.

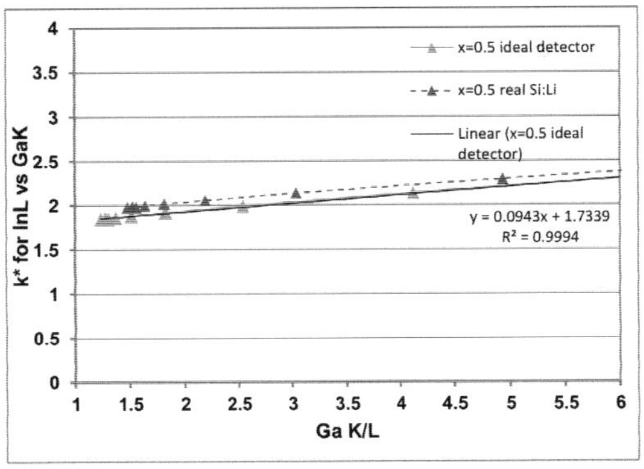

Figure 2b. CASINO simulations for of $k^*_{InL,GaK}$ for ideal and Si:Li detector with modelled ratios of detector efficiencies of $\varepsilon_{GaK}/\varepsilon_{GaL}$=1.196 and $\varepsilon_{GaK}/\varepsilon_{InL}$=1.072.

4. Experimental

As a test case, X-ray spectra were recorded from several regions of nine different specimens of InGaN thin layers of differing indium concentrations, deposited on GaN buffers and Al_2O_3 corundum substrates. The experiments were performed using a JEOL 2010F field-emission TEM operated at 197kV and equipped with an Oxford Instruments Si:Li detector with ultrathin Moxtek polymer window and 25° nominal take-off angle. Standard quantification procedures in the Oxford Instruments ISIS300 software yielded inconsistent values using the nominal thin film k-factors for Ga L and Ga K lines (diamonds in figure 3), even after approximate inclusion of absorption corrections (triangles).

Our new quantification scheme suggests indium concentrations are actually higher than the nominal concentrations for low x samples, and quantifications from Ga L and Ga K lines now agree very well for each spectrum. For the averages of our 9 specimens we get as difference between quantification using the Ga L and the Ga K line intensities as reference: $\Delta x = x_{k^*{}_{\mathrm{InL,GaL}}} - x_{k^*{}_{\mathrm{InL,GaK}}} = 0.005 \pm 0.007 \leq 0.01$. The error bars of the data in figure 3, where we plot the indium concentration determined using different EDXS method vs the nominal indium concentrations, as given by the scatter from each individual specimen, are significantly larger than above Δx because most samples showed a systematic decrease in apparent x with thickness, indicating indium depleted surface regions. Whether this apparently reduced indium content in the top of the layers is to be attributed to growth, or to argon ion milling, is presently unclear. However, we have excluded beam damage as a cause of this as we tested with a focused electron probe of same intensity that beam damage in the form of preferential loss of indium occurs only after a dose typically factors 20-50 above that used for the measurements shown in figure 3 [7]. For the eight specimens of higher indium concentrations measured in figure 3 it can be seen that the EDXS results from our k^* method lie between those values reported by the ISIS software without any absorption corrections using either Ga L or K lines but are generally closer to those results for Ga L than with Ga K quantification, so neither simple averaging of reported ISIS values nor absorption correction in ISIS will give similar improvements. The first data points for the measurement of the sample with the lowest indium concentration show a different trend, where ISIS and our values have negligible statistical error bars but do not overlap. We are presently testing whether this apparent outlier is due to chemical inhomogeneity of this specific sample.

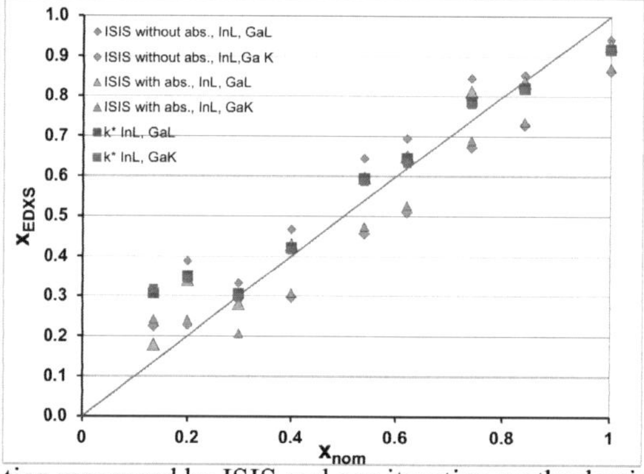

Figure 3: Indium concentration measured by ISIS and our iterative method using k^* values from figure 1.

5. Summary

A new self-consistent absorption correction has been tested on a series of $In_xGa_{1-x}N$ thin layers of various concentrations x, for which standard quantification procedures yielded inconsistent values using nominal k-factors for Ga L and Ga K lines, even after inclusion of absorption corrections. Our new quantification scheme suggests indium concentrations are actually higher than the nominal concentrations for low x samples. Quantifications from Ga L and Ga K lines now agree to typically better than $\Delta x \leq 0.01$=1at% for each spectrum. Total errors are larger because our samples showed an apparent decrease in x with sample thickness, indicating indium depleted surface regions.

References

[1] Cliff G and Lorimer GW 1975 *J. Microsc.* **103**, 203
[2] Morris PL, Ball MD, Statham PJ 1980, *Proc. EMAG'79*, Brighton, *Inst. Phys. Conf. Ser.* **52**, 413
[3] Watanabe M, Horita Z and Nemoto M 1996 *Ultramicroscopy* **65**, 187
[4] Walther T 2010, *Proc. EMAG 2009*, Sheffield, *J. Phys. Conf. Ser.* **241**, 012016
[5] Qiu Y, Nguyen VH, Dobbie A, Myronov M and Walther T 2013 *Proc. 18th Microsc. Semiconducting Mater. Conf.*, Oxford, *J. Phys. Conf. Ser.* **471**, 012031
[6] Parri MC, Qiu Y and Walther T (2014) submitted to *J. Microscopy*
[7] Walther T and Wang X (2015) *J. Microscopy,* in print with doi: 10.1111/jmi.12291
[8] Hovington P, Drouin D and Gauvin R 1997 *Scanning* **19**, 1

Radial composition variations in the shells of GaAs/AlGaAs core-shell nanowires

J S Nilsen[1], J F Reinertsen[2], A Mosberg[1], V T Fauske[1], A M Munshi[3], D L Dheeraj[3], B O Fimland[2,3], H Weman[2,3] and A T J van Helvoort[1]

[1]Department of Physics & [2]Department of Electronics and Telecommunications, Norwegian University of Science and Technology (NTNU), NO-7491, Trondheim, Norway.
[3]CrayoNano AS, Otto Nielsens vei 12, NO-7052, Trondheim, Norway.

E-mail: julie.s.nilsen@ntnu.no

Abstract. GaAs nanowires (NWs) are seen as promising building blocks for future optoelectronic devices. To ensure reproducible properties, a high NW uniformity is required. Here, a substantial number of both position-controlled and randomly grown self-catalyzed GaAs/AlGaAs core-shell NWs are compared. Single NWs are characterized by correlated micro-photoluminescence (μ-PL) spectroscopy and transmission electron microscopy (TEM). TEM is done in the ⟨110⟩- and ⟨112⟩-projections, and on the ⟨111⟩-cross-section of the NWs. The position-control grown NWs showed a higher degree of uniformity in morphology. All NWs on both samples had a predominantly stacking fault free zinc blende structure, with a main optical response around the GaAs free exciton energy. However, NW-to-NW structural variations in the tip region and radial compositional variations in the shell are present in both samples. These structural features could be the origin of variations in the optical response just below and above the free exciton energy. This correlated study demonstrates that the observed distinct, sharp PL peaks in the 1.6 - 1.8 eV energy range present in several NWs, are possibly related to radial compositional variations in the AlGaAs shell rather than the structural defects in the tip region.

1. Introduction

For the development of future optoelectronic devices based on NWs, self-catalyzed GaAs/AlGaAs core-shell NWs are an attractive model system [1]. The NWs are easily grown using a vapor-liquid-solid (VLS) mechanism, even on different types of lattice mismatched substrates due to the strain relaxation on the NW surface. However, the correlation between structural defects and the optoelectronic properties is still debated. One reason for this is that the NWs are subject to large NW-to-NW variations within a growth batch. Therefore, a significant sampling set combined with correlated optical and structural analysis of the same single NWs is required. To reduce the NW-to-NW variations for fundamental studies and more importantly for the smooth integration into larger scale devices, position-controlled growth has attracted quite some attention recently [2, 3]. Here we compare the structure of random and positioned-control grown NWs.

2. Experimental

The NWs in this study are grown on a Si(111) substrate both randomly and with position-control. The latter is achieved by covering the substrate with a 40 nm thick thermal SiO_2 film. Holes in the film about 100 nm in diameter and 1 μm apart were made by nanoimprint lithography [3]. The GaAs cores are grown through a self-catalyzed VLS mechanism using molecular beam epitaxy. Growth parameters for the two samples can be found in table 1.

Table 1. Growth parameters for the two samples. Shell growth is initiated before Ga droplet consumption and goes via a vapor-solid mechanism. Notation: Random/positon-controlled.

	Ga [ML/s]	As [Torr]	Al [ML/s]	[min]	[°C]
GaAs core	0.7/0.6	$8.0 \times 10^{-6}/5.5 \times 10^{-6}$	-/-	20/25	630
$Al_{0.33}Ga_{0.67}As$ shell	0.2/0.3	$9.0 \times 10^{-6}/1.0 \times 10^{-5}$	0.10/0.15	30/20	630
GaAs cap	0.2/0.3	$9.0 \times 10^{-6}/1.0 \times 10^{-5}$	-/-	15/8	630

For characterization, the NWs were dispersed on a 50 nm thick SiN support. The study was performed in a correlated manner, meaning the same individual NWs were studied first in μ-PL at 12 K and subsequently by TEM [4]. For each growth batch at least 15 NWs were studied. In addition to conventional TEM techniques, selected NWs were studied in the ⟨112⟩-projection as well as in the ⟨111⟩-cross-section (the latter made using focused ion beam (FIB)) using high angular annular dark field scanning TEM (HAADF STEM).

3. Results and discussion

Scanning electron microscopy (SEM) of as-grown samples (fig. 1(a, d)) illustrates the difference in morphology between random and position-controlled growth of NWs.

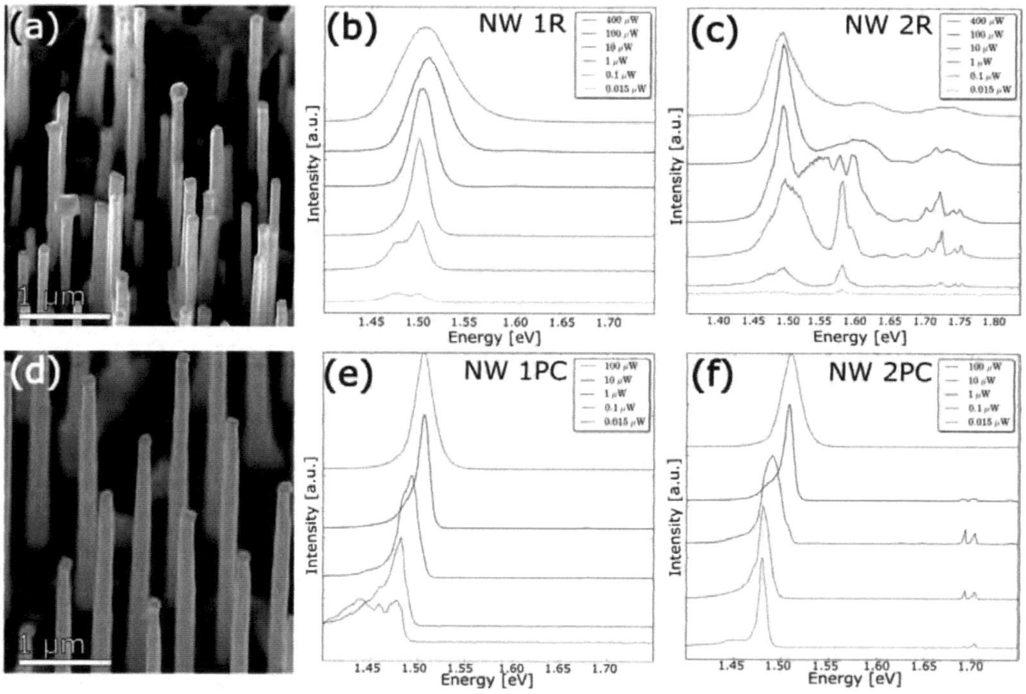

Figure 1. (a) and (d): Tilted SEM images of as-grown samples. Power-dependent PL spectra at 12 K are seen in (b, c) for two randomly and (e, f) for two position-control grown NWs.

The spacing between the NWs in the randomly grown sample is varying, giving local variations in the growth conditions for the cores and for the different facets during shell and cap growth.

The PL spectra show that there are differences in the optical properties between the two samples, but more pronounced are the NW-to-NW differences within each sample. PL spectra from four NWs are presented in fig. 1(b, c) and (e, f). All the spectra have a peak around 1.51 eV, assigned to the recombination of the GaAs free exciton [5]. The shoulder peak just below this energy seen for some NWs is most likely related to different types of defects [6]. The presence of peaks with energy above that of the free exciton (fig. 1(c, f)) are uncommon and previously not seen when the shell is thin and grown at lower temperature. These peaks are present in about 30% of the position-control grown NWs and in 70% of the randomly grown NWs studied. The energy range in which these peaks are found is 1.6 - 1.8 eV. To better understand the origin of this high-energy emission, structural characterization by TEM of the same NWs was performed. Stacked dark-field (DF) TEM images in the ⟨110⟩-projection from the tip region of the NWs in fig. 1(b, c) and (e, f) are shown in fig. 2(a-d), respectively.

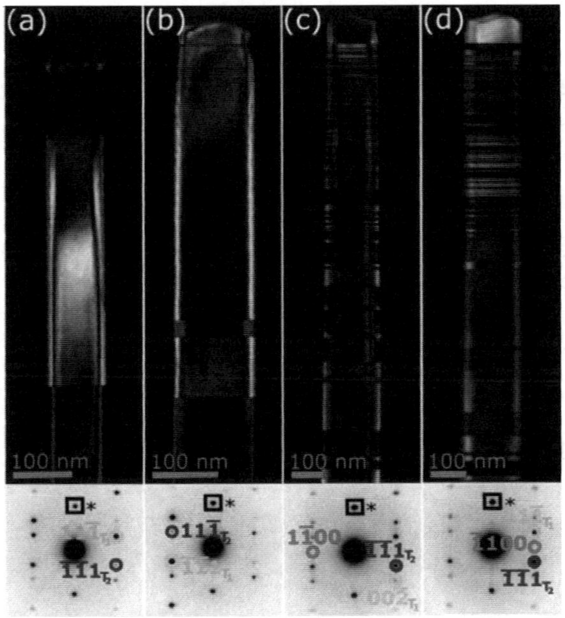

Figure 2. Stacked DF TEM images with the corresponding inverted diffraction patterns of the top part of the same NWs as characterized by PL. (a) NW 1R, (b) NW 2R, (c) NW 1PC and (d) NW 2PC. $[111_{T_1,T_2}]$ and WZ [0002] are marked with *.

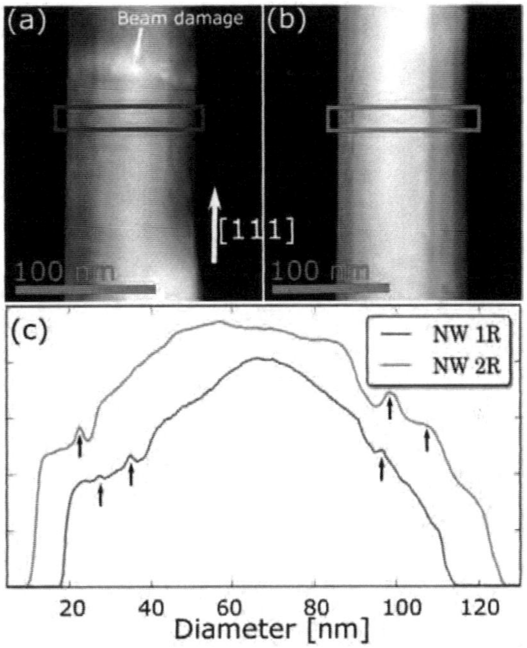

Figure 3. HAADF STEM images of (a) NW 1R and (b) NW 2R, taken in ⟨112⟩. (c) Intensity profiles from the areas marked in (a) and (b). Arrows mark peaks in HAADF STEM intensity.

For the NWs shown here, the position-control grown NWs have considerably more lattice defects than the randomly grown NWs, despite the more homogeneous morphology (fig. 1). On the other hand, the randomly grown sample has far more NW-to-NW variations in the morphology and defect structure considering the whole set of NWs studied. Most NWs in both samples had some wurtzite (WZ) phase present, which forms during Ga droplet consumption [7]. However, no relation between high-energy peaks and lattice defects in this correlated study could be identified. HAADF STEM characterization of several NWs tilted onto a ⟨112⟩-direction, revealed that the NWs had a line structure in the AlGaAs shell running along the length of the NWs (fig. 3). To study the shell structure more closely, cross-sectional FIB samples were prepared from the middle of these NWs, and their HAADF STEM images are given in fig. 4.

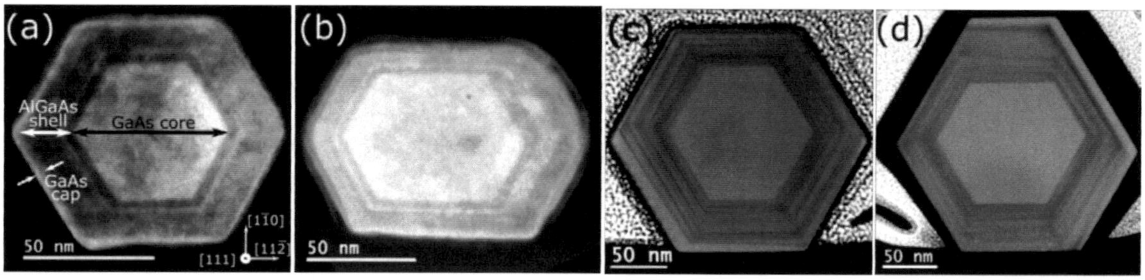

Figure 4. (a-d) HAADF STEM of the cross-sections of the NWs in fig. 2(a-d), respectively.

The cross-sections with an even specimen thickness revealed the presence of compositional variations in all six segments of the AlGaAs shell, matching the line structure in the HAADF STEM images of whole NWs in the [21$\bar{1}$]-projection (fig. 3). These lines are Al-deficient and Al-rich radial bands, and they vary in thickness both within and between each shell segment. Such bands are not commonly seen in this type of core-shell NWs. Instead, Al-enrichment in the ⟨112⟩-direction has been reported in NWs grown at low temperature shell growth [8]. It is possible that in some shell segments, the formation of an Al-deficient layer between two Al-rich barriers could lead to confinement of carriers, which could explain the high-energy features seen in the PL spectra of several NWs. However, the energies are higher than one would expect assuming a pure GaAs well, considering the thickness of the Al-deficient layers [9] .

4. Conclusions

By using correlated PL-TEM on a relatively large number of randomly and position-control grown GaAs/AlGaAs core-shell NWs, it is demonstrated that both growth methods have pronounced NW-to-NW variations within one batch. In addition to conventional TEM to study lattice defects, HAADF STEM in two different directions (⟨112⟩ and ⟨111⟩) on the same NWs showed the presence of fine-spaced radial compositional variation along the entire AlGaAs shell on both samples. These variations are different for every shell segment. Further work is needed to confirm if these radial compositional variations could be the sole explanation for the observed sharp high-energy (1.6 - 1.8 eV) PL emission present in several of the here studied NWs.

Acknowledgments

The Research Council of Norway is acknowledged for the support to the NorFab (197411/V30) and the NORTEM (197405) facilities, as well funding the projects 239206/O70 and 214235/F20.

References

[1] Fontcuberta i Morral A 2011 *IEEE J. Sel. Top. Quant.* **17** 819
[2] Gibson S J, Boulanger J P and LaPierre R R 2013 *Semicond. Sci. Tech.* **28** 105025
[3] Munshi A M *et al.* 2014 *Nano Lett.* **14** 960–966
[4] Todorovic J, Moses A, Karlberg T, Olk P, Dheeraj D L, Fimland B O, Weman H and van Helvoort A T J 2011 *Nanotechnology* **22** 325707
[5] Ahtapodov L *et al.* 2012 *Nano Lett.* **12** 6090–6095
[6] Spirkoska D *et al.* 2009 *Phys. Rev. B* **80** 245325
[7] Krogstrup P, Curiotto S, Johnson E, Aagesen M, Nygård J and Chatain D 2011 *Phys. Rev. Lett.* **106** 125505
[8] Kauko H, Zheng C L, Zhu Y, Glanvill S, Dwyer C, Munshi A M, Fimland B O, van Helvoort A T J and Etheridge J 2013 *Appl. Phys. Lett.* **103** 232111
[9] Shi T *et al.* 2015 *Nano Lett.* **15** 1876–1882

Characterisation of TiC layers deposited using an electrical discharge coating process

S J Algodi[1], J W Murray[2], A T Clare[2] and P D Brown[1]

[1]Division of Materials, Mechanics and Structures, and [2]Institute of Advanced Manufacturing, Department of Mechanical, Materials and Manufacturing Engineering, Faculty of Engineering, University of Nottingham, University Park, Nottingham, NG7 2RD, UK.

Abstract. Electrical discharge machining (EDM) is a non-conventional, high-accuracy machining process for the manufacture of complex shapes, regardless of hardness of the workpiece. There is interest to develop the EDM technique for coating or surface modification by using a powder metallurgy (PM) tool electrode and/or added powder suspended within the dielectric fluid. We report on the EDM deposition of TiC coatings onto stainless steel, using either Cu or TiC electrodes, with and without Ti powder in the working oil. EDM processed layers exhibited hardness values ~ 3-4 times higher than the substrate, emphasising the ability of EDM to impart improved mechanical performance to the surface of austenitic stainless steel.

1. Introduction

Titanium carbide (TiC) is the most widely used ceramic carbide protective coating owing to its high melting point, hardness, oxidation resistance and chemical stability, along with excellent wear resistance and low coefficient of friction [1]. High quality TiC coatings have been deposited using methods such as CVD, plasma spraying, sputtering and ion plating. An advantage of EDM is that it provides for localised coating or surface modification, without need for complicated equipment. Also it can be applied to complex shapes [2,3]. In this context, there is a need to appraise the process-structure-property interrelationships of EDM since this technique is still at the developmental stage.

2. Experimental

It is not easy to improve the wear resistance of austenitic stainless steel because they cannot be hardened by heat treatment. For this work, commercial 304 stainless steel substrates were used, cut into 20 x 20 x 6 mm sections and polished to a mirror finish. The coating work was performed using a die sinking machine (Mitsubishi EA12V; current 10 A; gap voltage 320 V; negative tool polarity; on-time 8μs; off-time 256 μs; machining time 10 minutes). Two types of tool electrode were compared, *i.e.* a powder metallurgy (PM) TiC electrode and a conventional solid Cu electrode, with and without Ti powder (45 μm) dispersed into the dielectric fluid (hydrocarbon oil) contained within a stirrer tank (figure 1). The workpiece was fixed to the bottom of the tank, while the tip of the electrode was set to touch the workpiece top surface. The electrode moved towards the workpiece under servo-control and the gap voltage was allowed to rise until it became sufficient to ionise the dielectric, with short duration discharges in the gap between the two electrodes, leading to material replacement or alloying on the localised scale. Metallurgical analyses of the workpiece surfaces and the coatings in cross-section was performed using scanning electron microscopy (SEM), energy dispersive X-ray spectroscopy (EDS), X-ray diffractometry (XRD) and microhardness testing.

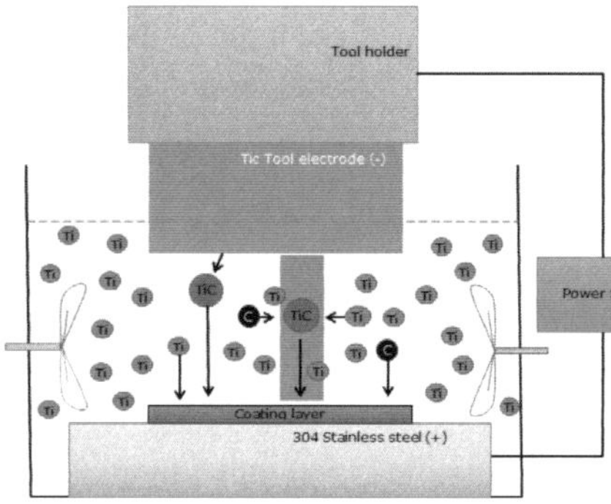

Figure 1. The electrical discharge coating (EDC) process.

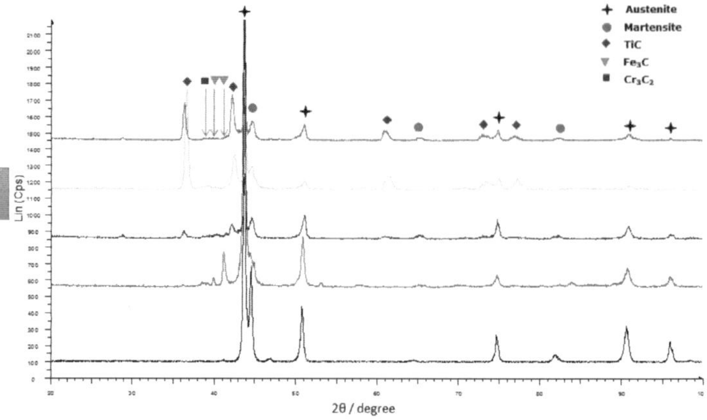

Figure 3. XRD patterns for: a) 304 stainless steel; and EDM surface treated samples processed using: b) Cu electrode; c) Cu electrode & Ti powder; d) TiC electrode; and e) TiC electrode & Ti powder.

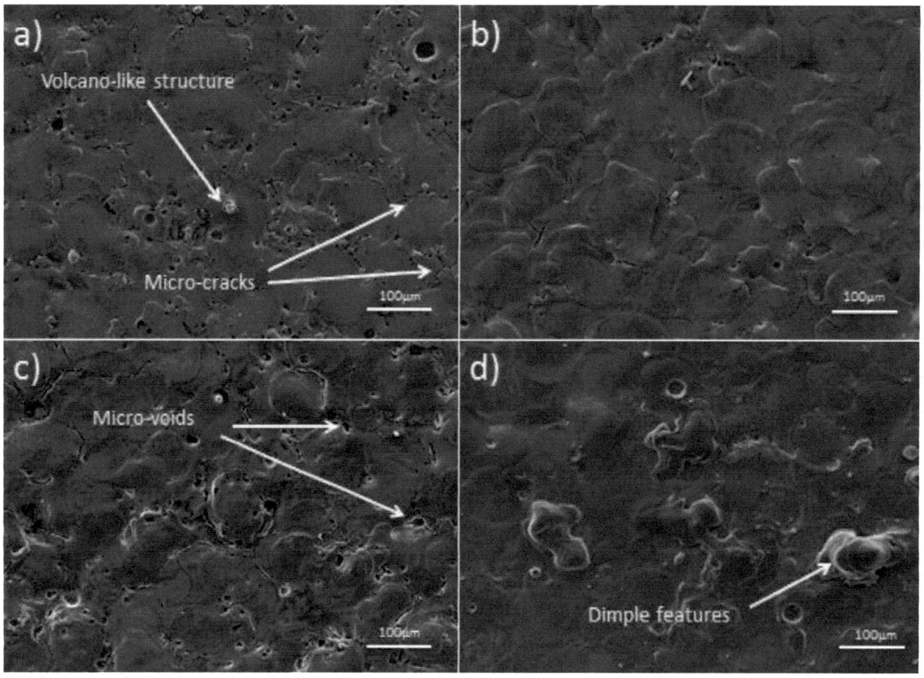

Figure 2. SE images of EDM processed 304 stainless steel surfaces using: a) Cu electrode; b) Cu electrode & Ti powder; c) TiC electrode; and d) TiC electrode & Ti powder.

3. Results & Discussion

Figures 2(a-d) present secondary electron (SE) micrographs of EDM processed samples, prepared with Cu and TiC electrodes, with and without Ti powder. The surfaces produced using the conventional Cu electrode, without Ti powder, exhibited uniform cracking, voids and 'volcano-like' structures (figure 2a), attributable to high thermal energy followed by rapid cooling during discharges. With the addition of Ti powder, near surface porosity and cracking were almost eliminated (figure 2b). Individual and overlapping craters could be distinguished, indicative of less intensive melting, with previous craters left unaffected, noting that the addition of powder to the dielectric fluid resulted in reduced resistance, increased gap distance and consequent reduction in electrical discharge power density [1].

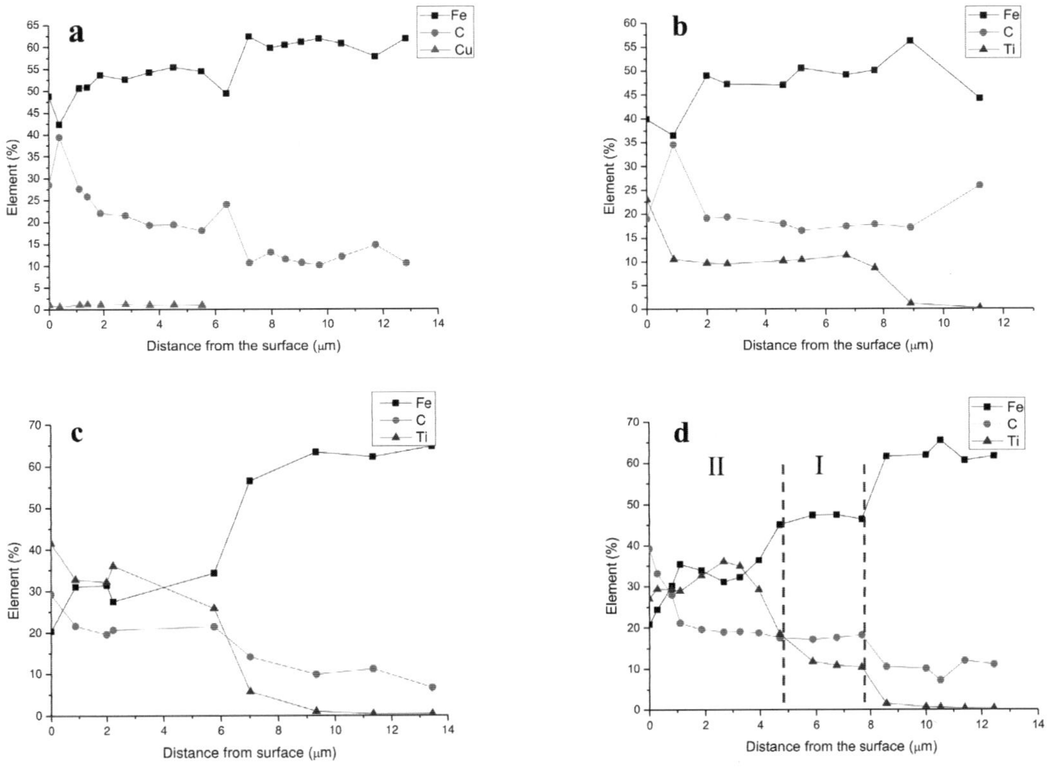

Figure 4. EDS line profiles for EDM processed layers in cross-section, on 304 stainless steel substrates, fabricated with: a) Cu electrode; b) Cu electrode & Ti powder; c) TiC electrode; and d) TiC electrode & Ti powder.

Cracking and porosity were more prevalent on the surface produced using a TiC PM electrode (figure 2c). The porosity may be attributed to the expulsion of gas bubbles from molten material during solidification [4]. It is noted that the melting point of TiC (3143 K) is slightly higher than the boiling point of Fe (3023 K), so when the TiC layer started to solidify, Fe would still be molten, and cracks would tend to develop within the TiC layer upon gradual cool-down due to differential thermal contraction [5]. The addition of Ti powder to the dielectric combined with the use of a TiC electrode led to the production of smoother, denser coatings than those produced with a Cu electrode (figure 2d). In particular, differently shaped dimple features were present which were found to be rich in Ti (figure 2d). This effect was attributed to the movement of Ti powder to form clusters between the two electrodes leading to a localized reduction in the dielectric fluid resistance, promoting discharge, along with the breaking of clusters of TiC powder from the PM electrode which deposited on the workpiece surface, forming dimpled shapes [6]. A ~ 50% reduction in mean crack density was determined for the Cu electrode / Ti powder samples, as compared with samples prepared using the Cu electrode only. Similarly, a ~ 30% reduction in crack density was observed for the TiC electrode / Ti powder samples, as compared with coatings produced using the TiC electrode only.

The crystal structures of the EDM processed samples were appraised using XRD (figure 3). For coatings processed with the Cu electrode, the evidence showed the formation of Fe_3C due to reaction between the substrate and the hydrocarbon of the dielectric fluid (figure 3b), whilst addition of Ti powder led to the formation of TiC and the inhibition of Fe_3C (figure 3c). Surfaces prepared with the TiC electrode exhibited TiC rich phases (figure 3d). EDS analyses were performed to appraise the chemical compositions of the processed coatings in cross-section geometry. Significant amounts of C were associated with surfaces prepared using the Cu electrode, again consistent with the formation of Fe_3C. EDS line profiles suggested that these surface modified layers were ~ 5.5 μm in depth, below which Cu was not detected (figure 4a). The addition of Ti powder led to the development of a uniform

TiC layer ~ 8 μm in thickness (figure 4b) with consequent inhibition of Fe_3C formation in keeping with the XRD data. Much higher levels of Ti were found for EDM samples processed using the TiC electrode (figure 4c) whilst the addition of Ti powder led to the development of a banded profile, indicating initial reaction dominated by the powder in the dielectric fluid, prior to deposition from the TiC electrode (figure 4d, I and II, respectively).

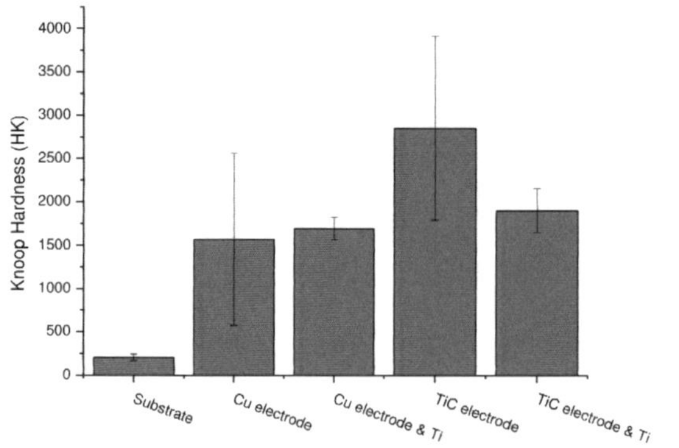

Figure 5. Knoop hardness values for the EDM processed layers.

Knoop hardness measurements were performed in cross-section on all samples using 20 indents for each and with a low force of 10 g (figure 5). It was evident that the EDM processed layers exhibited hardness values ~ 3-4 times higher [2,7] than that of the substrate, emphasising the ability of the EDM process to impart improved mechanical performance to the surfaces of austenitic stainless steel. Large variations in indent size were observed for all samples, attributable to local variations in composition and microstructure. The coatings produced using TiC electrodes without Ti powder showed the highest mean values of hardness of ~ 2800 HK (figure 5), consistent with the enhanced XRD TiC profile of figure 3d, as compared to figure 3e, suggesting denser TiC formation. The increasing hardness in the surface modified layers processed with the Cu electrode, with and without Ti powder in the dielectric fluid, was attributed to the development of TiC and Fe_3C, respectively. Further enhancement of hardness in the surface modified layers produced using the TiC electrode was achieved in the presence of Ti powder, but maximised in the absence of Ti powder, again suggesting that the electrode contributed higher quality TiC into the surface modified layer. The EDS line profile transition indicated in Figure 4d suggested local depletion of Ti powder, with TiC from the electrode dominating the latter stages of deposition.

4. Summary

The novel EDM technique as used to deposit TiC coatings on stainless steel substrates has been appraised. Structure determination confirmed the presence of TiC within the recast layer, deposited using a TiC electrode. The addition of Ti powder to the dielectric oil had significant impact on the coating process with the production of smoother surfaces free of cracking and porosity. However, coatings with highest levels of hardness were produced in the absence of Ti powder.

Acknowledgements

The authors would like to thank Nigel Neate and Martin Roe for their invaluable technical support; and the Ministry of Higher Education, Iraq, for supporting the lead author.

References

[1] Yih-Fong T and Fu-Chen C 2005 *Journal of Materials Processing Technology* **170** 385-391
[2] Wang ZL, *et al.* 2002 *Journal of Materials Processing Technology* **129** 139-142
[3] Akiyoshi M *et al.* 2003 *JSME International Journal Series A* **46** 496-501
[4] Chen S.-L. *et al.* 2014 *Applied Surface Science* **311** 47-53
[5] Furutani K, Sato H and Suzuki M 2009 *The International Journal of Advanced Manufacturing Technology* **40** 1093-1101
[6] Suzuki T and Kobayashi S 2013 *Electrochimica Acta* **114** 844-850
[7] Xiao H *et al.* 2014 *Surface and Coatings Technology* **258** 1006-1010

The Structure-Phase Compositions of Powder Ni – based Coatings after Modification by DC Plasma Jet Irradiation

D Alontseva[1] and E Ghassemieh[2]
[1]*East-Kazakhstan State Technical University,*
69 Protozanov St, Ust-Kamenogorsk, 070004, Kazakhstan
[2]*School of Mechanical and Aerospace Engineering, Queen's University of Belfast,*
Stranmillis road, Belfast, BT9 5AH, UK

E-mail: [1]dalontseva@mail.ru, [2]e.ghassemieh@qub.ac.uk

Abstract. This paper presents the results of investigation of the structure-phase compositions of Ni-based coatings deposited by plasma jet on steel substrates after modification by direct current (DC) plasma jet irradiation. Transmission Electron Microscopy (TEM), Scanning Electron Microscopy (SEM), Atomic Force Microscopy (AFM) and X-ray Diffraction (XRD) are used in the current study. The phase structures and morphology of precipitation of strengthening phases from solid solution are defined. The irradiation of the coatings leads to the evolution of the structural-phase state of coatings: an increase in the volume fraction of hardening intermetallic phases, the formation of sufficiently homogeneous fine-grained structure in the irradiated coatings. There is a mutual penetration of the substrate main element Fe into the coating and base coating elements Ni into the substrate as a result of the coating treatment by a pulse DC plasma jet.

1. Introduction

Surface modification by plasma jet has replaced other thermal treatments due to the short and efficient exposure time requirement [1,2] . The main problems of plasma sprayed protective coatings are their porosity, lack of homogeneity on account of poor agglomeration of powder particles, and high surface roughness. These result in insufficient corrosion and wear resistance of such coatings. To eliminate these drawbacks the coatings may be re-treated by application of electron beam, laser beam, or plasma jet [1,2]. In the current study, a novel choice of the DC plasma re-treatment modes has been used. This includes optimum choice of the power density of a plasma source on the coating surface and also an efficient speed of the source movement. A specific temperature field in the system of coating-substrate has been created. This is to provide the precipitation of hardening phases in the coating during the added treatment. Through this procedure the surface of the coating is melted, whereas the area of the interface of the coating with substrate is exposed to temperature of about 400°C for several minutes. In our previous study the temperature distribution in two-layer absorbers are calculated using both permanent linear case model and variable thermal parameters non-linear model [3,4]. On the basis of the model calculations we recommended the optimum irradiation modes. These modes ensure

[1] Address for correspondence: D Alontseva, East-Kazakhstan State Technical University, 69 Protozanov St, Ust-Kamenogorsk, 070004, Kazakhstan. E-mail: dalontseva@mail.ru

temperature profiles which results in accelerated of diffusion process. Also they assist with precipitation of hardening intermetallic phases in protective powder coatings.

The aims of the current research is to establish the regularities of the evolution of structural and phase composition and properties of Ni based powder coatings following additional treatment by DC pulse plasma jet. The behavior of these materials under irradiation will subsequently be experimentally verified.

2. Materials and Methods

Protective coatings of 100 -200 μm thickness were deposited on steel substrate. These were further processed by the "Impulse-6" plasma detonation unit. The substrate material was 10^4 μm thick steel 3 (Fe – base, C - 0.25 %, Mn – 0.8 %, Si – 0.37 %, P < 0.045 %). PG-19N-01 Ni- based powders alloy with additives of Cr (8…14%), B (2.3%), Si (1.2…3.2%), Fe (5%), C (0.5%) were used for coating purpose. The powder fractions varied in size from 50 to 80 μm. The powder coatings were deposited in air. The modes of deposition of these coatings were described in detail in our previous papers [3,4]. The irradiation of the samples by DC plasma jet was carried out at power density of $1.9 \cdot 10^9$ W/m^2 and travel speed of 0.006 m/s.

Experimental methods of analysis included Transmission Electron Microscopy (TEM) (JEM-2100 ("JEOL", Japan) and TECNAI, (Phillips)), X-ray diffraction (XRD) (X'Pert PRO ("PANalytical", the Netherlands)), Scanning Election Microscopy (SEM) (JSM-6390LV ("JEOL", Japan) with EDS ("Oxford Instruments", Great Britain)) and Atomic Force Microscopy (AFM) (NT-206 (Belorussia)). M-691 Precision Ion Polishing System ("Gatan", USA) was used to prepare TEM foils by the Ar ion sputter etching method. Microhardness test of the samples was performed using a digital microhardness tester (LM-700 (LECO, Russia)).

3. Results and Discussion

The untreated coatings showed a high surface roughness with Ra of approximately 70 nm (Fig.1a). They also displayed a sharp, structurally heterogeneous boundary with the substrate (Fig. 2a). According to the XRD data the PG-19N-01 coating before modification contains a Ni-based fcc-lattice phase with the parameter of a=3.525…3.540 Å. This is in accordance with the estimated electron-diffraction pattern of 3.53 Å.

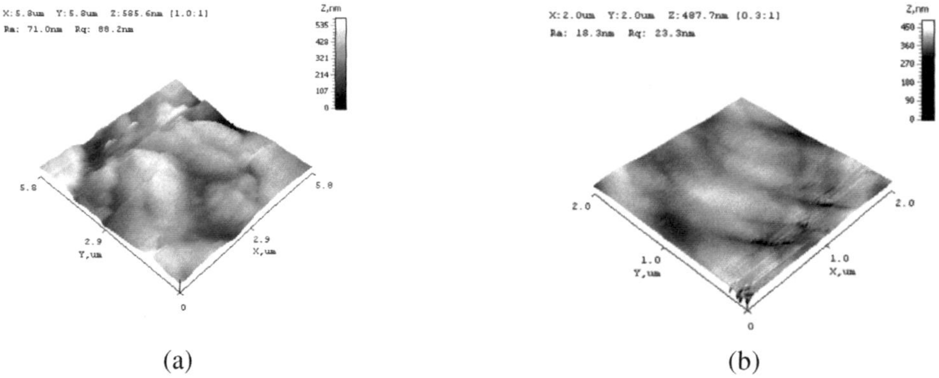

(a) (b)

Figure 1. AFM-images of the surface of PG-19N-01 coatings before (a) and after (b) irradiation by DC Plasma Jet

TEM results indicate that the base through-thickness layer of coatings is a mixture of crystallographically differently oriented nanograins of the fcc-matrix with the size of 1-2 nanometers, and lamellas of intermetallic phases up to 50 nanometers long (Fig. 3a). The indexed reflex of intermetallic phases (Fig. 3b) was tested by the dark field method (Fig. 3c).

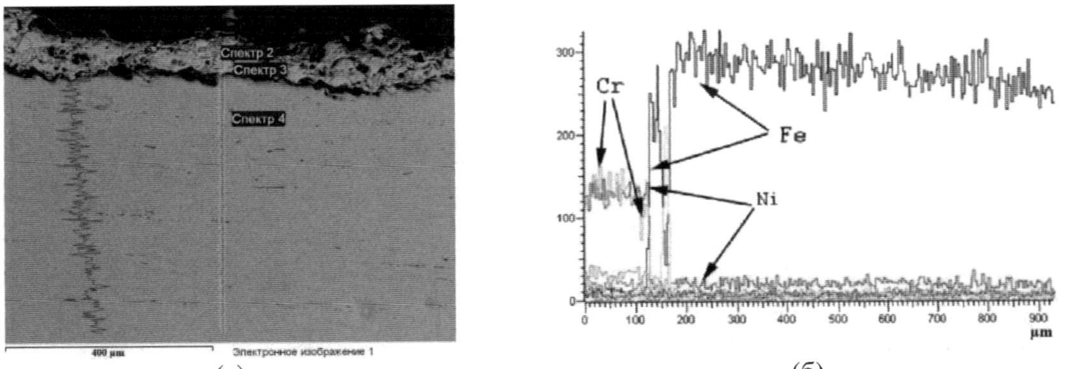

(a) (б)

Figure 2. SEM - image of the non-modified sample cross section (the coating at the top of the image) with a line of spectrum (a) and the corresponding distribution of elements (b)

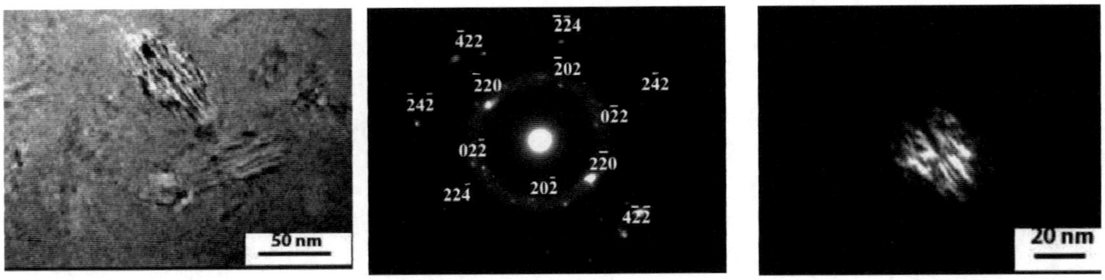

(a) (b) (c)

Figure 3. TEM - images of PG-19N-01 coating: the $CrNi_3$ particle (bright field) (a); the electron diffraction pattern of a $CrNi_3$ particle with the zone axis [111] (b); the $CrNi_3$ particle (dark field) shot in point reflex ($\bar{2}20$) (c)

The type of crystal lattice and the interplane distances matches the experimentally observed ones for $CrNi_3$. According to the database of "2011 International Centre for Data Diffraction" (the PDF Card No 01-071-7595) for the $CrNi_3$ phase (Chromium Nickel, Cubic, Fm-3m, 225), its crystal lattice parameter is 3.552 Å. In the current coating, the estimated fcc-lattice parameter of $CrNi_3$ is a= 3.6 Å (fig 3b). This is very close to the data obtained by XRD which is (a=3.555 Å). The TEM images indicate that the volume ratio of $CrNi_3$ in the coating PG-19N-01 before irradiation is about 20%. These nanosize intermetallic phases are highly strengthening. It has been shown that the greatest microhardness of the coating occurs at locations of high volume concentration of these phases [3,4]. The maximum microhardness of PG-19N-01 before modification is 6.0 GPa. The microhardness at some distance from the surface shows that the coating has considerably higher hardness compared to the steel substrate. The average microhardness of the steel is about 1.4 GPa. The microstructure of the coating in this case could be compared to composite material with nanosized reinforcement.

After modification by DC plasma jet the roughness decreases in average by 3.5 times (from Ra= 70 nm to Ra= 20 nm). This is due to the melting of the surfaces of coatings through additional treatment (Fig 1 b). TEM and XRD examinations reveal that the volume concentration of the $CrNi_3$ phase in the coatings is increased to about 25%. This is about 5% higher than the untreated coatings. Correspondingly, an increase of about 1GPa in microhardness was observed. At the same time, additional processing does not affect the particle size, particle morphology, and the crystal lattice parameter (Fig.4 c, d). The matrix structure of the coating is also unaffected by irradiation (Fig.4 a, b). However, the microstructure of the coatings after irradiation has changed to a fine-grained homogeneous microstructure (Fig.5 a). Also SEM with EDS indicates more than 100 microns penetration of Ni from the irradiated Ni-based coatings into the substrate (Fig. 5b). Also penetration of Fe from the substrate into the irradiated coatings could be clearly observed (Fig 5c). Both observations suggest the development of radiation-enhanced diffusion in the absorbing materials during irradiation.

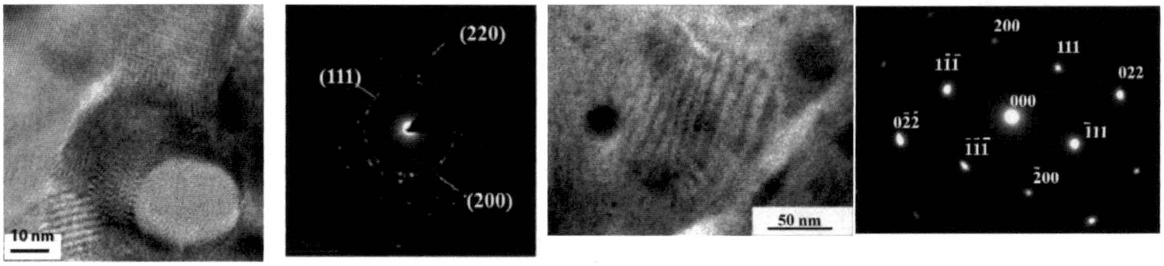

(a) (b) (c) (d)

Figure 4. TEM images of the Ni-based irradiated coatings (a), corresponding diffraction pattern (b) the $CrNi_3$ particle (c) and its diffraction pattern (d)

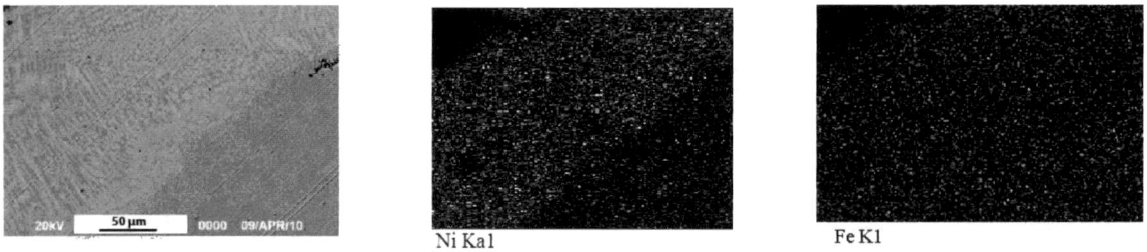

(a) (b) (c)

Figure 5. SEM image of modified Ni-based coating (the left part of the image) on steel substrate (the right part of the image) (a), corresponding map of Ni distribution (b) and the map of Fe distribution (c)

4. Conclusions

The irradiation of the coatings by a DC pulse plasma jet according to the recommendations of model calculations modes leads to the evolution of the structural-phase state of coatings: This results in an average of 5% increase of the volume fraction of hardening intermetallic phases. It also leads to the formation of a sufficiently homogeneous fine-grained structure in the irradiated coatings. The strengthening particles of intermetallic phases are precipitated in the form of nanosized lamellae. SEM observations show redistribution of the elements of the coating into the substrate. This indicates the acceleration of diffusion processes during irradiation.

5. Acknowledgment

The authors thankfully acknowledge the funding provided by the Scientific Committee of the Ministry of Education and Science of the Republic of Kazakhstan for the project No 29-417-15 titled “Automated precision deposition of powder coating multifunctional systems”.

6. References

[1] Davis J 2004 *Handbook of Thermal Spray Technology* (Ohio: ASM International, Materials Park) p 248

[2] Kadyrzhanov K 2005 *Ion-beam and Ion-plasma Modification of Materials* (Moscow: Moscow State University) p 212

[3] Alontseva D, Krasavin A, Pogrebnjak A, Russakova A. 2013 Modification of Ni-based plasma detonation coatings by a low-energy DC E-Beam *Acta Physica Polonica A* 123(5) 867 – 870

[4] Alontseva D, Pogrebnjak A, Kolesnikova T, Russakova A 2013 Modeling of processes in Co-based coatings exposed to plasma jet irradiation *Medziagotyra* 19 (3) 277 – 282

Microstructure characterization of hypereutectoid aluminium bronze composite coating

P Kucita[1], S C Wang[1], W S Li[2], R B Cook[1] and M J Starink[3]

[1]national Centre for Advanced Tribology at Southampton (nCATS), Faculty of Engineering and the Environment, University of Southampton, University Road, Southampton, SO17 1BJ, United Kingdom
[2]State Key Laboratory for Advanced Non-ferrous Materials, Lanzhou University of Technology, China
[3]Engineering Materials, Faculty of Engineering and the Environment, University of Southampton, University Road, Southampton, SO17 1BJ, United Kingdom

Email: pk4v07@soton.ac.uk

Abstract. Hypereutectoid aluminium bronze coating was deposited onto an E.N. 10503 steel substrate using plasma transferred arc welding (PTA). Microstructure characterisation of the coating and a section near the steel substrate joint was carried out using SEM, EBSD, EDS in conjunction with XRD and depth-sensing nano-indentation. The constituent phases in the coating were identified as: martensitic Cu_3Al β_1' phase, solid solution of Al in Cu α phase and the intermetallic Fe_3Al κ_1 phase. The region near the steel substrate was characterised by high hardness, large grains and presence of Cu precipitates. No cracks were observed in this region. The coating has high hardness of 4.9GPa and Young's modulus of 121.7GPa. This is attributed to homogeneous distribution of sub microns size Fe_3Al intermetallic phase. The implications of the coating to the engineering application of sheet metal forming are discussed.

1. Introduction

Hypereutectoid Cu-Al-Fe aluminium bronzes are known to have high hardness and wear resistance. However, production of these alloys using conventional sand casting is difficult due to the complications associated with the narrow range of solidification and the eutectoid reaction which can lead to embrittlement[1]. These manufacturing difficulties can be overcome by using rapid solidification techniques such as plasma transferred arc welding (PTA). The high cooling rate can lead to novel properties due to a microstructure that deviates substantially from thermodynamic equilibrium. The microstructure of Cu-Al-Fe alloy is diverse and highly dependent on the cooling rate. An ordering reaction[2], martensitic transformation[3] and formation of intermetallic have been reported. This makes the microstructure highly complex. Several phases are known to exist; these are listed in table 1. The microstructures of as-cast and heat treated Cu-Al-Fe alloys are well studied. However, there are few reports on the phases present in Cu-Al-Fe coatings, particularly those with high Al content. In the present study, the microstructure and mechanical properties of a Cu-11Al-7Fe (in wt.%) aluminium bronze coating produced by PTA was assessed using SEM, EBSD, EDS, XRD and depth-sensing nano-indentation.

Table 1: Possible phases in rapidly solidified Cu-Al-Fe alloy in the composition range 9-14 wt.% Al

Phase	Composition	Crystal structure	Reference
β' disordered martensite	Cu_3Al	Cubic	[2]
β_1' ordered martensite	Cu_3Al	Orthorhombic a=0.449nm, b=0.519nm, c=3.820nm	[4]
Υ' ordered martensite	Cu_3Al	Hexagonal a= 0.2606nm, c/a = 1.619	[5]
α solid solution of Al in Cu	(Cu)	Cubic a =0.361nm	[6]
Υ_1 intermetallic	Cu_9Al_4	Cubic a =0.871nm	[6]
κ_1 intermetallic	Fe_3Al	Cubic a=0.571nm	[7]
κ_2 intermetallic	FeAl	Cubic a=0.290nm	[7]

2. Experimental

2.1. Materials

The hypereutectoid aluminium bronze powders were produced from component metals with an electrolytic purity of 99.98% by gas atomisation in nitrogen atmosphere and cooling water as detailed in reference[8]. The powders were then deposited onto an E.N. 10503 medium carbon steel by plasma transferred arc welding (PTA) using a current of 170A, 3-6l min^{-1} plasma nitrogen flow rate, 1-3g s^{-1} powder feed rate and stand-off distance of 5-10mm. The composition of the powder and coating are shown in Table 2.

Table 2: Composition of the hypereutectoid powders and coating (all in wt.%)

	Cu	Al	Fe	Mn	Ni	Si
Powder	78.9	14.9	4.4	1.4	0.5	0.3
Coating	78.9	11.2	6.9	0.9	0.3	0.2

2.2. Approach to characterisation

Specimens for characterisation were sectioned using electric discharge machining. Samples were ground up to 4000 grit SiC and either diamond polished or electro-polished, the former by using 1μm diamond polishing and the latter by electrolysing for ~15s in 20% HNO_3, 15% 2-butooxyethanol and 65% methanol solution (in volume) cooled to -30°C at 30V. All specimens for EBSD investigation were prepared by electro-polishing. EBSD mapping was carried out using a JEOL JSM-6500F FEG-SEM operating at an accelerating voltage of 15 keV and probe current of 0.1nA. The specimen was tilted to 60° with a working distance of 18mm. Step size of 0.5μm and 1μm were employed. Phase identification was performed using a Bruker D2 Benchtop X-ray diffraction machine. The hardness and Young's modulus were evaluated using depth-sensing nano-indentation operating at 500nm depth control mode. A total of 196 indents were produced

3. Results and discussion

An SEM image in backscattered electron (BSE) mode of an electro-polished specimen is shown in Figure 1(a). Precipitates (appearing dark) with a range of dimensions from sub-microns to tens of microns were clearly observed. Certain areas of the matrix were also selectively removed following the electro-polishing (appearing dark grey in Figure 1a). This suggests the presence of multiple phases in the matrix. Table 3 shows typical EDS analysis of the precipitates and the constituent phases in the matrix. The precipitates are rich in Fe and Al with stoichiometric ratio close to 3:1. In conjunction with XRD, the precipitates were identified as the κ_1 Fe_3Al phase. EDS analysis of the unaffected area of the matrix shows that it is high in Cu and Al. The stoichiometric ratio of Cu to Al suggests that it is one of the martensitic phases. As shown in table 1 several martensitic phases have been reported. The disordered martensite is known to form at Al<11 wt.%[2], which falls outside the composition of the coating. For alloys with Al>11wt.% an ordering reaction precedes the martensitic transformation into either β_1', β_1' + Υ' or Υ'[2][5]. The ordered Υ' with hexagonal structure has been reported to form at 12.9<Al<14.7 wt.%[5].

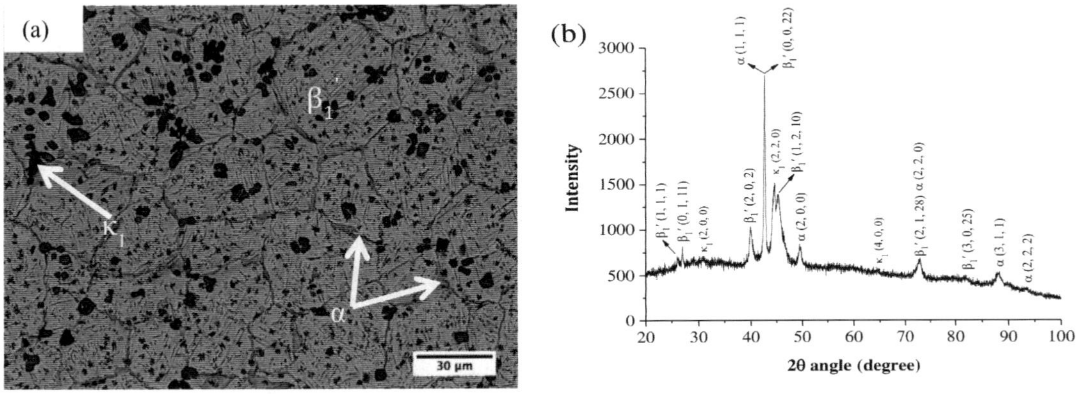

Figure 1 (a) BSE image of the electro-polished specimen, (b) observed XRD spectrum.

Based on this information and the observed XRD spectrum, the martensitic phase in the present coating was identified as the ordered β_1' with orthorhombic structure. The structure identified is in agreement with previous TEM observation of water quenched Cu-11.7Alwt.% alloy[4]. In summary, using XRD, SEM and EDS, the phases in PTA Cu-11Al-7Fe coating were identified as: martensitic Cu_3Al β_1' phase, solid solution of Al in Cu α phase and the intermetallic Fe_3Al κ_1 phase.

Table 3: EDS analysis of the precipitates and the constituent phases in the matrix (all in at.%)

Phases	Cu	Al	Fe	Mn	Ni	Structure
α (Cu)	79.1± 0.6	17.8±0.3	2.1±0.3	0.6±0.1	0.4±0.2	Cubic, cF4 a=b=c=0.365nm
β_1' (Cu_3Al)	74.4±0.5	22.6±0.6	1.5±0.4	1.0±0.1	0.5±0.2	Orthorhombic, oP8, a=0.449nm, b=0.519nm, c=4.66nm
κ_1 (Fe_3Al)	11.6±1.8	24.5±1.2	57.9±2.1	1.5±0.2		Cubic, cF16 a=b=c=0.579nm

Figure 2(a) shows the EBSD phase map of the coating. As shown in the magnified area, EBSD is capable of identifying the κ_1 and the α phase but it was not possible to index the β_1' phase.

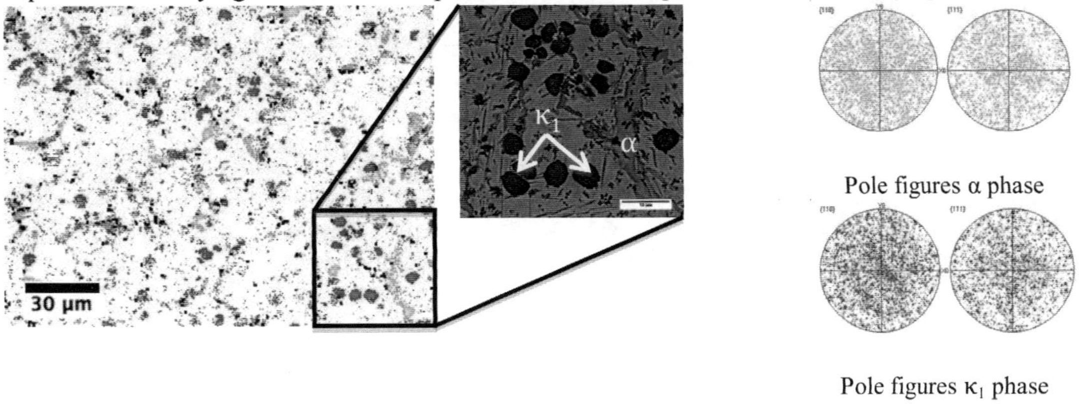

(a) (b)

Figure 2 (a) EBSD phase map (red= κ_1 phase, green = α phase, white = not indexed) and BSE image of the magnified area, (b) corresponding pole figures of the EBSD phase map.

Asymmetric broadening of the β_1' phase peak was observed as shown in figure 1(b). This indicates crystallite smallness, presence of long-range internal stresses and chemical heterogeneity. Fine precipitates also contribute to broadening of the observed XRD peaks[10]. These factors affect the crystal structure, which can impede indexing. The pole figures and misorientation angle distribution indicate that there is no crystallographic texture. This is expected, as the turbulence generated by the

PTA process will result in a random orientation of grain growth. Fine precipitates (<1 μm) were also observed in BSE image shown in Figure 2(a). EDS analysis indicates that it is rich in Fe and Al with composition close to that of the κ_I phase. The homogeneous distribution of the κ_I phase and the predominantly hard martensitic β_1' phase leads to a uniform hardness with an average of 4.9GPa and Young's modulus of 121.7GPa. The martensitic structure such as the one found in this coating was reported to have higher wear resistance than a similar alloy with eutectoid structure[9]. Previous work[8] has also reported that due to the limited solubility of Cu and Al in Fe, the adherence tendency of Cu-Al-Fe alloys to steel is low. This makes the coating highly suitable as forming die materials. Figure 3(a) shows the EBSD orientation map of the joint between coating and the steel substrate. This area is characterised by large grains of ~17μm in size. In this area the κ_I phase and the grains in the joints were also observed to have the same orientation. This suggests that the region melted during deposition is thicker than the joint region. Dendrites were observed to extend from the joint region into the coating as shown in Figure 3(b). This increases the surface area of contact between the coating and the substrate, which increases the bond strength.

(a) (b)

Figure 3: (a) EBSD orientation map, (b) BSE image of the joint

The joint area is rich in Fe with significant amount of Al and traces of Cu as shown in Table 4. This suggests that during the deposition process part of the steel substrate was melted. For PTA process, melting of the substrate is not uncommon as droplet temperature can exceed 2000°C, which is well above the melting point of the steel substrate. The melting of the steel substrate is evidenced as an increase in the Fe content in the coating compared to the original powder as shown in Table 2 as well as the presence of large grains in the joint region as shown in Figure 3(a). The presence of Cu and Al in the interface layer leads to a hardness of 282HV compared to 195HV for the steel substrate. No cracks were observed in this region.

Table 4: Compositions of the joint area between the coating and steel substrate (all in at.%)

Cu	Al	Fe	Mn	Ni	Si	Cr
0.7±0.2	6.0±0.8	88.2±1.5	1.1±0.13	0.5±0.1	1.3±0.2	2.3±0.2

4. Conclusion

Microstructure characterisation of Cu-11Al-7Fe (in wt.%) coating identified the presence of 3 main phases, martensitic β_1' phase, solid solution of Al in Cu α phase, and Fe_3Al κ_I intermetallic phase. The presence of the martensitic and intermetallic phases leads to high hardness of 4.9GPa. The high hardness of the coating is expected to improve its wear performance, thus making it suitable for applications requiring wear resistance such as die for sheet metal forming.

References

[1] Alam S, Sasaki S and Shimura H 2001 *Wear.* **248** 75–81
[2] Kwarciak H, Bojarski J and Morawiec Z 1986 *J. Mater. Sci.* **21** 788–792
[3] Murray JL 1985 *Int Met Rev.* **30** 211–233
[4] Zenji K and Setsuo N 1963, *Jpn. J. Appl. Phys.* **2** 478–486
[5] Nakanishi N 1960 *Trans. JIM.* **2** 79–86
[6] Ponweiser N, Lengauer CL and Richter KW 2011 *Intermetallics.* **19** 1737–46
[7] Li WS, Liu Y, Wang Z, Ma C and Wang S 2012 *Trans. Nonferrous Met. Soc. China.* **22,** 2139–45
[8] Li WS, Wang Z, Lu Y, Gao Y and Xu JI 2005 *Trans. Nonferrous Met. Soc. China.* **16** 607–612
[9] Shi Z, Sun Y, Bloyce A and Bell T 1996 *Wear.* **193** 235–241
[10] Ungár T 2004 *Scripta Materialia.* **51** 777-781

Cross sectional TEM analysis of duplex HIPIMS and DC magnetron sputtered Mo and W doped carbon coatings

J Sharp[1*], I Castillo Muller[1], P Mandal[2], A Abbas[3], G West[3], W M Rainforth[1], A Ehiasarian[2], P Hovsepian[2]

1: Dept. of Materials Science and Engineering, University of Sheffield, Sir Robert Hadfield Building, Mappin St., Sheffield S1 3JD, UK
2: Materials and Engineering Research Institute, Sheffield Hallam University, City Campus, Howard Street, Sheffield, S1 1WB, UK
3: Department of Materials, Loughborough University, Loughborough, LE11 3TU, UK

*j.sharp@sheffield.ac.uk

Abstract. A FIB lift-out sample was made from a wear-resistant carbon coating deposited by high power impulse magnetron sputtering (HIPIMS) with Mo and W. TEM analysis found columnar grains extending the whole ~1800 nm thick film. Within the grains, the carbon was found to be organised into clusters showing some onion-like structure, with amorphous material between them; energy dispersive X-ray spectroscopy (EDS) found these clusters to be Mo- and W-rich in a later, thinner sample of the same material. Electron energy-loss spectroscopy (EELS) showed no difference in C-K edge, implying the bonding type to be the same in cluster and matrix. These clusters were arranged into stripes parallel to the film plane, of spacing 7-8 nm; there was a modulation in spacing between clusters within these stripes that produced a second, coarser set of striations of spacing ~37 nm.

1. Introduction

Carbon based coatings provide excellent resistance to mechanical wear for their substrate, and are therefore of great interest for use on components that are required to resist a harsh working environment in critical applications. These coatings are deposited by various techniques such as pulsed closed-field unbalanced magnetron sputtering (CFUBMS), cathodic arc evaporation, magnetron sputtering and plasma assisted chemical vapour deposition (1). Recently a new technique, High Power Impulse Magnetron Sputtering, has been introduced, providing highly ionised metal plasma for deposition of highly dense and adherent coatings (2). The paper discusses the fine microstructure of Me-doped Carbon coatings deposited by a combined HIPIMS-DC Magnetron sputtering process, where in this case Me = (Mo, W). The advantages of this combined process are high rate deposition and precise control over the coating structure, texture and residual stress (3).

2. Methods

Mo and W doped carbon coatings were deposited onto steel by non- reactive mixed process of DC magnetron sputtering and HIPIMS (2) using an industrial size HAUZER 1000- 4 coating system. The substrates were rotated inside a sputtering chamber with four magnetron sources furnished by sputtering targets of pure graphite and Me. Prior to the coating deposition the substrate was bombarded by accelerated Me ions generated by the HIPIMS discharge, which provides for high coating adhesion (scratch adhesion critical load values, Lc=70 N). This procedure was followed by the deposition of 120 nm MeN transition layer followed by 1.8 μm Me-doped carbon layer.

Cross-sectional TEM samples were prepared from the coatings using the FEI Nova dual beam FIB-SEM at Loughborough University. TEM characterisation was initially carried out in a JEOL 2010F TEM (200 kV) at Sheffield University, and further analysis carried out in the JEOL double aberration corrected R005 (300kV), both with energy dispersive X-ray spectroscopy (EDS), electron energy-loss spectroscopy (EELS) and scanning TEM (STEM) capability.

3. Results and discussion

Low magnification STEM and CTEM images (Figure 1) show the ~1800 nm thick film to have multiple levels of striated structure: a set of finer stripes of spacing 7-8 nm wide, and a coarser set of spacing ~37 nm wide. Contrast from these regimes of stripes was complimentary between HAADF and BF STEM images.

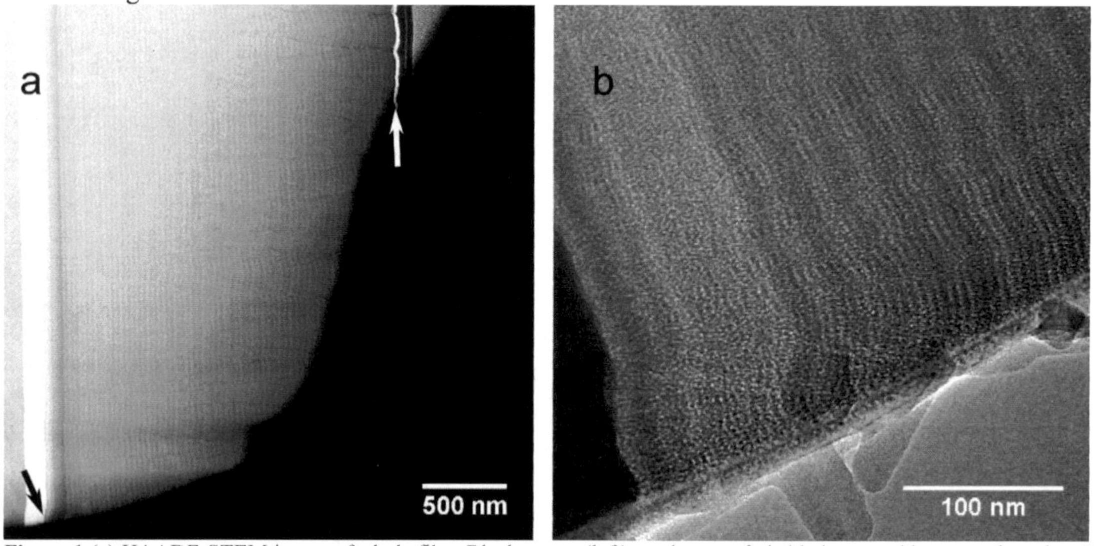

Figure 1 (a) HAADF-STEM image of whole film. Black arrow (left) marks top of nitride layer and bottom of coating. White arrow (right) marks top of film and gold protective coating deposited before FIB. (b) CTEM image of bottom of film showing two regimes of striations on different scales.

The finest set of striations are seen at higher resolution to be made of onion-like clusters, shown in Figure 2. EDS spot microanalysis showed very small concentrations of Mo and W in both clusters and matrix, with little discernible difference between cluster and matrix. EDS spot analysis of a thinner sample (0.05 inelastic mean free paths) from a wear test prepared later showed more Mo and W in the clusters than the matrix, in areas far from the wear surface which should be comparable to this unworn sample. STEM contrast supported this, with the clusters appearing brighter in HAADF and darker in BF. EELS showed no difference in the C-K edge between cluster and matrix, implying the bonding type to be the same in both. Unfortunately, though high-resolution HAADF-STEM was possible at the very edge of the specimen, the disordered nature of the surrounding material did not allow for good enough contrast to determine the positions of the metal atoms inside the clusters as yet.

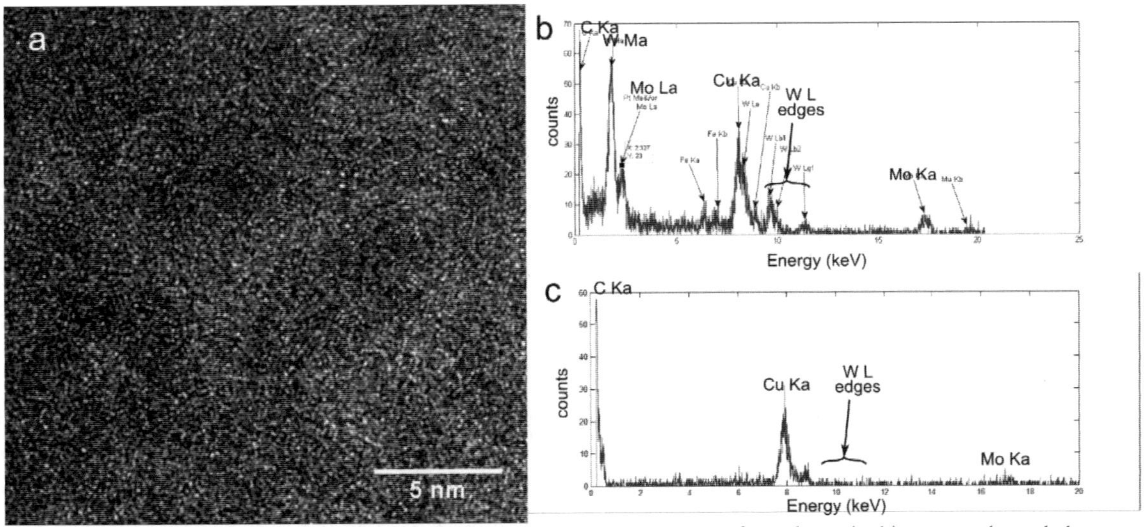

Figure 2 (a) HRTEM image showing onion-like parts in clusters. (b) EDS spectrum from cluster in thinner sample made later (not this sample, but deposited by the same procedure). (c) EDS from the matrix in this sample.

The finer stripes are made of these clusters arranged in lines. The coarser stripes appear to be a modulation in the spacing of the fine stripes and in the intensity from the matrix material between them (Figure 3); EDS from wide areas in lighter and darker stripes show the HAADF-STEM-brighter stripes to contain more Mo and W.

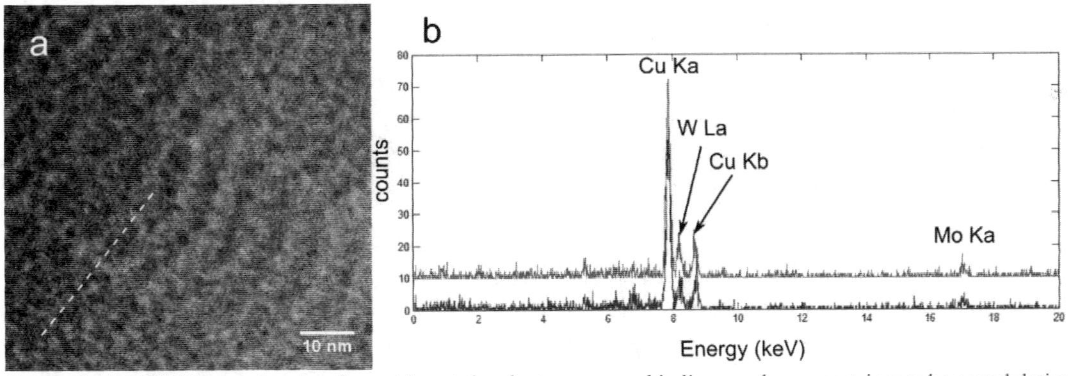

Figure 3 (a) CTEM image showing finer stripes to be clusters arranged in lines, and coarser stripes to be a modulation in the inter-cluster spacing within these lines. The top-left of the image is in a darker stripe; the bottom-right is in a lighter stripe, with the dotted line marking the angle of the stripe. (b) EDS spectra from wide areas within a HAADF-lighter coarse stripe (green top line) and within a HAADF-darker coarse stripe (blue bottom line). The HAADF-light stripe has a little more Mo and W.

The striated microstructure exists within columnar grains that persist up through the coating as it grows. The grain boundaries are denuded of clusters so are seen as dark in HAADF STEM (Figure 4) and bright in CTEM and BF STEM. The stripes in the microstructure are domed across the grain, indicating the boundaries have an effect on the process of clusters organising into stripes.

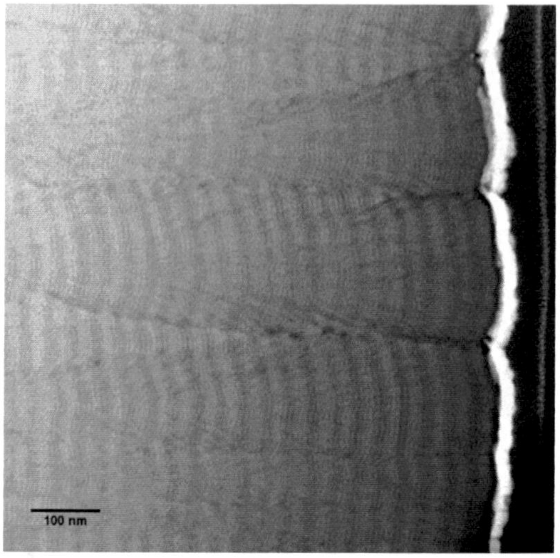

Figure 4 HAADF STEM image of the top of the coating showing columnar grains with domed tops (white line is gold protective coating applied before starting FIB).

The coating has been found to be extremely wear-resistant in the wear test work which is ongoing as the next phase of this project; the changes in microstructure which achieve this high wear resistance are complex and not yet fully characterised.

4. Conclusions

This HIPIMS carbon film co-deposited with Mo and W has a columnar microstructure that persists through the thickness of the film. Within these columnar grains, the material is organised into clusters with an onion-like structure observable in some places; EDS microanalysis shows these clusters to contain more of the Mo and W, while EELS analysis shows no change in the C-K edge between cluster and matrix. The clusters are organised into lines, giving fine stripes parallel to the film plane, 7-8 nm thick, proceeding up the grains. Within the stripes, there is a further modulation in the spacing between the clusters, giving a wider set of stripes of spacing ~37 nm.

Further work will characterize the microstructures developed under different deposition conditions (in particular the incident energy imparted to the atoms during deposition), and the microstructural changes that occur at the surface during wear.

References

[1] Voevodin AA, Rebholz C, Schneider JM, Stevenson P, Matthews A 1995 *Surface and Coatings Technology* vol 73 pp 185-197

[2] Ehiasarian AP 2008 *Fundamentals and Applications of High Power Impulse Magnetron Sputtering* in: *Plasma Surface Engineering Research and its Practical Applications,* ed. Ronghua Wei (Trivandrum, India: Research Signpost)

[3] Hovsepian P, Arunachalamsugumaran A, Purandare Y, Loch D, Ehiasarian A 2014 *Thin solid films* vol 562 pp 132-139

STEM electron tomography in the Scanning Electron Microscope

M Ferroni[1,3], A Signoroni[1], A Sanzogni[1], G Sberveglieri[1,3], A Migliori[2], L Ortolani[2], M Christian[2,4], L Masini[2], V Morandi[2]

[1] Department of Information Engineering, Brescia University, Via Valotti 9, 25123 Brescia, Italy

[2] CNR-IMM Section of Bologna, via Gobetti 101, 40129 Bologna, Italy

[3] CNR-INO, Via Valotti 9, 25123 Brescia, Italy

[4] CNR-ISOF, via Gobetti 101, 40129 Bologna, Italy

Corresponding author's email: matteo.ferroni@unibs.it

Abstract. The scanning-transmission imaging mode in the SEM allows for the three-dimensional tomographic reconstruction of a specimen, starting from a set of projection images. Compressed sensing was used to solve the undetermined problem of structure reconstruction and was proven capable of overcoming the limitations arising from the sampling scheme. Reconstructions of cobalt particles within a carbon nanotube and collagen fibrils in a dermal tissue are presented, demonstrating the potential of this technique in the set of 3-D electron microscopy methods for both physical and biological science.

1. Implementation of tomography in the SEM

In conventional three-dimensional imaging methods operated in the scanning electron microscope, the volume reconstruction of a sample is retrieved from a cumulative series of images of the specimen, which is sectioned stepwise by a microtome blade or a focused beam of Ga ions during the imaging sequence [1]. This *slice-and-view* approach is capable of achieving nanometric resolution in the reconstruction of volumes as large as several cubic micrometres; however there are intrinsic limitations related to dealing with porous structures, measuring an adequate image contrast for the sample details, and maintaining a planar thin sectioning throughout the slicing process.

Recently, implementation of the scanning-transmission (STEM) imaging mode in the SEM has allowed the investigation of the inner structure of thin samples at low beam energy [2-4], and established the possibility of reconstructing the spatial mass distribution, similar to the approach to electron tomography (ET) in the TEM [5]. Because this method requires a projection scheme, reconstruction of the specimen relies on the backprojection algorithm of tomography [6] and could take advantage of the theory of compressed sensing to reduce the artifacts produced by the inevitably limited and incomplete series of experimental projections [7].

In this paper, we present the experimental approach and application of this tomographic method implemented in the SEM.

2. Experimental set-up and reconstruction workflow

ET in the SEM requires that the specimen is suspended to allow transmission of the electrons, while rotating under the electron beam to produce the series of projection-images. A dedicated segmented detector, capable of *Bright/Dark Field* imaging, and a specimen holder have been set up for the purpose, an arrangement that benefits from the absence of post-specimen imaging lenses in the SEM chamber. Owing to the strong influence of the mass-thickness of the sample on the beam scattering angle, the assembly of the holder and electron detector is fundamental to preserve the monotonic increase of image contrast upon tilting the sample, which is required for the subsequent reconstruction.

The tomographic workflow, summarized in **Figure 1**, is based on the conventional inverse-Fourier backprojection and uses the prior knowledge about the sparsity of the object under investigation to refine the tomogram, according to the Compressed Sensing (CS) approach to signal reconstruction.

Owing to the linear orthographic nature of the projections and their alignment, the tomographic reconstruction of the volume $r(x,y,z)$ can be decomposed into a *slice-by-slice* problem from the stack of lines of the projection images. Projections are considered as samples of the 2D Fourier transform of the volume slice, and arranged to form a polar sampling (with a missing wedge in our case). From this starting point, the effective backprojection is carried out by the Non-Uniform FFT [8], which offers a useful initialization and effectively operates in a CS-based refinement.

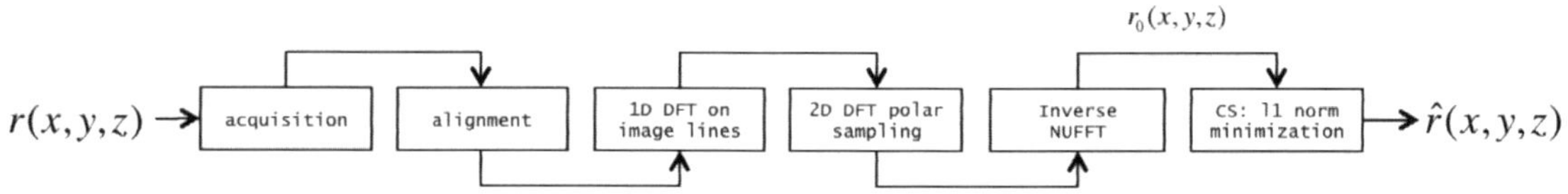

Figure 1 Schematic workflow of the reconstruction process.

According to the theory of CS applied to electron tomography, the initial tomogram $r_0(x,y,z)$ is refined by considering some degree of sparsity in both image and gradient domains [8]. Despite the presence of reconstruction artifacts caused by the limited number of experimental projections and the missing wedge, the method can converge to an accurate reconstruction $\hat{r}(x,y,z)$ and was here applied to samples with different characteristics, in order to test its feasibility in both physical and biological science. The regularization parameters have been optimized accordingly.

3. Application in materials science: carbon nanotubes and cobalt nanoparticles

Figure 2 (*Left*) shows a STEM image of a multi-walled carbon nanotube (CNT) filled with cobalt particles supported by a TEM holey carbon film grid.

A series of 53 projections was acquired by tilting the sample through -60° to +60° at 2° steps. The best image contrast for the visualization of the support, the CNT, and the Co particles was obtained for the *Dark-Field* imaging, where only the annular part of the detector was activated. The Region of Interest (ROI) selected for the reconstruction is marked by a red box, and the preliminary alignment of the tilt series was carried out by maximizing a rotation compensated image cross-correlation. The result of the reconstruction is presented in **Figure 2** (*Center and Right*). The lateral views clearly separate the Co particles within the light carbon structure. The reconstruction has been refined through compressive sensing with optimized regularization parameters for both the image and gradient domains [9]. In this case, the marked separation in the elemental composition and the localization of Co into particles account for the recorded predominant sparsity in the gradient domain.

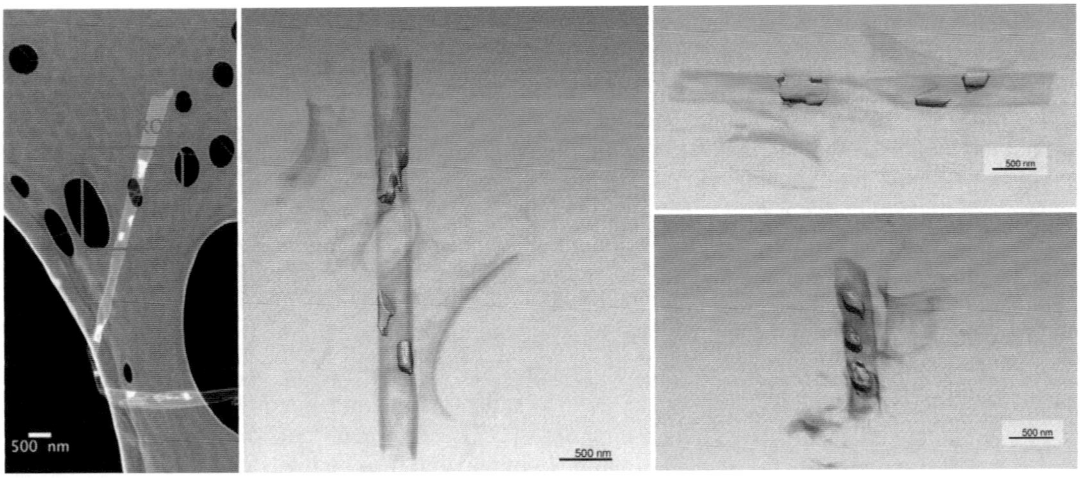

Figure 2 Tomographic reconstruction of a materials science sample. *(Left)* STEM image from the tilt series of a CNT filled with Co particles (*Center*) Front-view of the reconstructed tomogram (*Right*) Two side views of the tomogram showing the cobalt particles inside the CNT.

4. Application in biology: dermal tissue

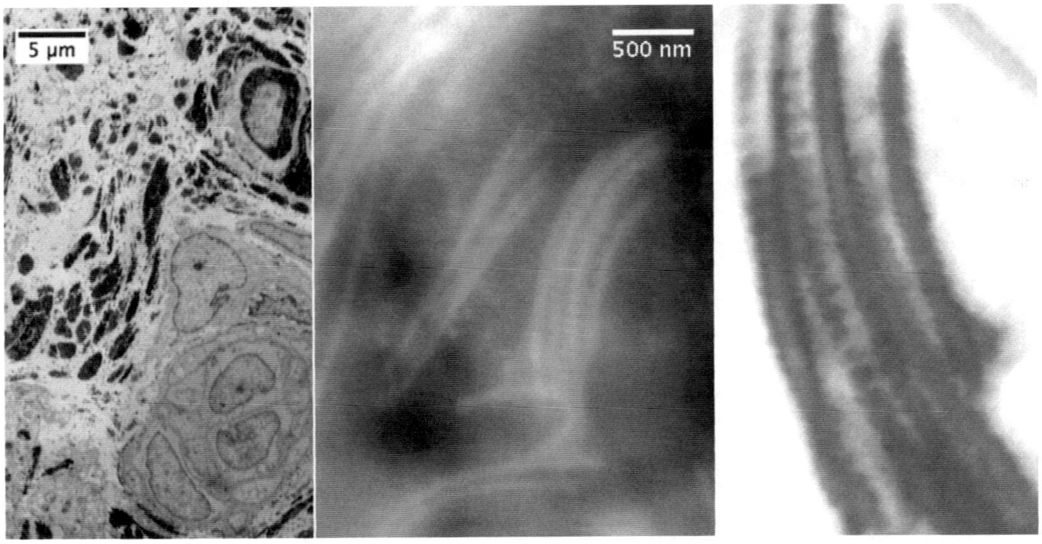

Figure 3 STEM images and reconstruction of a biological sample: (*Left*) Panoramic view of cellular structures and collagen fibers in human derma (*Center*) Image from the tilt series of the collagen bundle selected for tomographic reconstruction (*Right*) Detail from the CS-refined tomogram of the collagen bundle, showing the characteristic periodical (approx. 64 nm) striation along the fibrils.

Figure 3 highlights the structure of the dermal tissue as revealed by the STEM imaging mode in the SEM. The panoramic *Bright-Field* image of a 700 nm-thick specimen shows cellular membranes together with bundles of collagen fibrils. By operating the STEM in the *Dark-Field* mode with a 20 kV beam, the contrast for the fibrils is significant and adequate for the tomographic workflow.

In this sample, the region selected for the reconstruction encloses a small bundle of collagen fibrils. Starting from 91 projections covering the angular range between -50° and +40°, at 40000× magnification, the CS-refinement of the tomogram revealed the known transverse striation of the fibrils, which is visible as a periodical contrast modulation in the visualized tomogram. The resulting resolution was consistent with Crowther's estimation of 24 nm, which ideally requires the projections to be noise-free, perfectly aligned, and covering the entire angular range.

Owing to the complex characteristics of the biological system under investigation, optimization of the CS regularization parameters occurred for a dominant influence attributed to the sparsity term in the image domain.

5. Concluding remarks

The proposed approach exploits the flexibility of the SEM platform and showcases the potential of SEM tomography for the 3D analysis of volumes. This low-energy implementation of electron tomography, based on STEM imaging, is capable of reconstructing small samples and dealing with specimens featuring uneven thickness, or large contrast variations. The flexibility of optimization through compressed sensing allows the analysis of a large variety of samples.

The limit in resolution is determined by the probe size of the microscope, the specimen composition and thickness, and the number of projections that can be acquired without significant beam damage of the sample. The proposed system can be applied for both biological and physical science and performs at smaller volume and higher resolution ranges compared to conventional techniques.

References

[1] Cantoni M and Holzer L 2014 *MRS Bull* **39** 354–60
[2] Merli P, Morandi V and Corticelli F 2003 *Ultramicroscopy* **94** 89–98
[3] Morandi V, Merli P and Ferroni M 2006 *J Appl Phys* **99** 043512
[4] Morandi V, Merli P G and Quaglino D 2007 *Appl Phys Lett* **90** 163113
[5] Bals S, Van Aert S and Van Tendeloo G 2013 *Current Opinion in Solid State & Materials Science* **17** 107–14
[6] Kak A C and Slaney M 1999 *Principles of Computerized Tomography* (New York: IEEE press) pp 1–26
[7] Leary R, Saghi Z, Midgley P A and Holland D J 2013 *Ultramicroscopy* **131** 70–91
[8] Fessler J A and Sutton B P 2003 *Signal Processing, IEEE Transactions on Signal Processing* **51** 560–74
[9] Thomas J M, Leary R, Midgley P A and Holland D J 2013 *J Colloid and Interface Science* **392** 7–14

 doi:10.1088/1742-6596/644/1/012013

DIY Tomography sample holder

L Lari[1], I Wright and E D Boyes
The York-JEOL Nanocentre and Physics Department, The University of York, Heslington, YO10 5DD, York, UK

E-mail: leonardo.lari@york.ac.uk

Abstract. A very simple tomography sample holder at minimal cost was developed in-house. The holder is based on a JEOL single tilt fast exchange sample holder where its exchangeable tip was modified to allow high angle degree tilt. The shape of the tip was designed to retain mechanical stability while minimising the lateral size of the tip. The sample can be mounted on as for a standard 3mm Cu grids as well as semi-circular grids from FIB sample preparation. Applications of the holder on different sample systems are shown.

1. Introduction

Electron Tomography (ET) is a powerful tool for investigating 3-dimensional (3D) structures. Since its first application in 1968 [1], ET has been widely used in material science, life science and biology. This use boomed in the last decade due to availability of the modern computing resources for the 3D reconstruction. The technique is based on the acquisition of a series of images by varying the tilt angle of the sample respect to the beam [2]. Electron microscopy images are acquired in TEM mode mostly for biological samples and in annular dark field STEM for crystalline sample. A tilt series comprises typically of images taken every 1-2 degrees from at least -60 to +60 degrees. The manual acquisition of the series can be quite a lengthy exercise where the position of the region of interest of the sample has to be kept in the field of view in each frame. Stability of the goniometer stage is paramount to decrease the alignment time as well as for keeping the beam exposure at the minimum for electron beam sensitive samples. Recent development on the integration of tomography systems to the electron microscopes has brought software capable of automatically tracking the area of interest, focussing and tilting (*e.g.* FEI Xplore3D™). For the post acquisition analysis of datasets, software packages have been developed for 3D reconstruction are available freely (e.g. TOMOJ on FIJI [3]) or from companies providing ET dedicated holders.

Special holders for ET are in fact available from vendors with features including: 360 degrees on-axis rotation from needle like samples analysis to, dual axial holders, beryllium tip analytical holders, vacuum- and cryo-transfer holders. Most of these holders need a large pole piece gap > 5mm to allow space for axial rotation.

The York Jeol Nanocentre features a 2200FS JEOL aberration corrected microscope (imaging and probe corrected) modified with a large (4.3mm) pole piece gap [4] hosting a gas inlet system [5]. Although most of the above mentioned holders are designed for gaps above 5mm and they would not fit the York gap, it would certainly accommodate ET holders dedicated to High Resolution pole piece

[1] To whom any correspondence should be addressed.

gaps which are around 2.3mm. Encouraged by the space availability at the pole piece, we have found an economic solution by building in-house an ET system.

2. DIY ET holder

This project started with a purchasing investment (about two orders of magnitude cheaper than a full ET commercial solution), by acquiring from JEOL a standard tip plate for a single tilt holder. The holder tip was then machined into an arrow shape to minimise his hindrance in view of the tilt inside the pole piece gap. At the same time, enough material had to be retained to maintain its mechanical stability.

Figure 1. (Top) Cad drawing of the tip; **(Bottom left)** Overview of the sample holder tip; **(Bottom right)** Close up view with dimensions.

The sample holding brass *T*-piece was left completely unchanged. Two wings with a 20 degree slope were cut; this angle is larger than the one of the pole piece ensuring no contact with the pole piece cones during tilt. All the corners were smoothed with a fine file. The circular slot for the sample was left intact by the mechanical cut of the holder tip. The sample holder, in this way, can accommodate standard 3.05mm TEM sample grids as well as semi-circular Focused Ion Beam (FIB) grids. It also allows the sample to be rotated at will within its slot during mounting, and hence having a system compatible with the "dual-axis" tomography series acquisition, a procedure used widely to reduce the missing cone information problem [6].

3. Test and results

The holder was tested in a pole piece simulation chamber where a 4.3 mm pole piece gap has been replicated and the holder position can be studied as *in-situ*. Figure 2 shows the holder sideways and frontally at 0 degree position and at high tilt. Clearance from the simulated pole piece could be verified up to 90 degrees around the holder central axis. Within the microscope the holder was allowed to rotate within a +/- 70 degrees range only for safety. In fact once Z and X-Y stage movements are introduced, the maximum tilt angle is either reduced to +/- 60 degree range or the full Z-X-Y stage movement range in not available. This depends on the sample location on the TEM grid.

Figure 2. (Left) side view of the holder in the pole-piece replica chamber. **(Centre)** front view at 0 tilt. **(Right)** front view at 80 degree tilt.

A tilt series was acquired for cluster of TiO_2 nanopowder on a carbon film support as shown in Figure 3. High Angle Annular dark field images are shown for tilt angles from -60 to +50 tilt. The last image shows the grid bar starting to shadow the field of view.

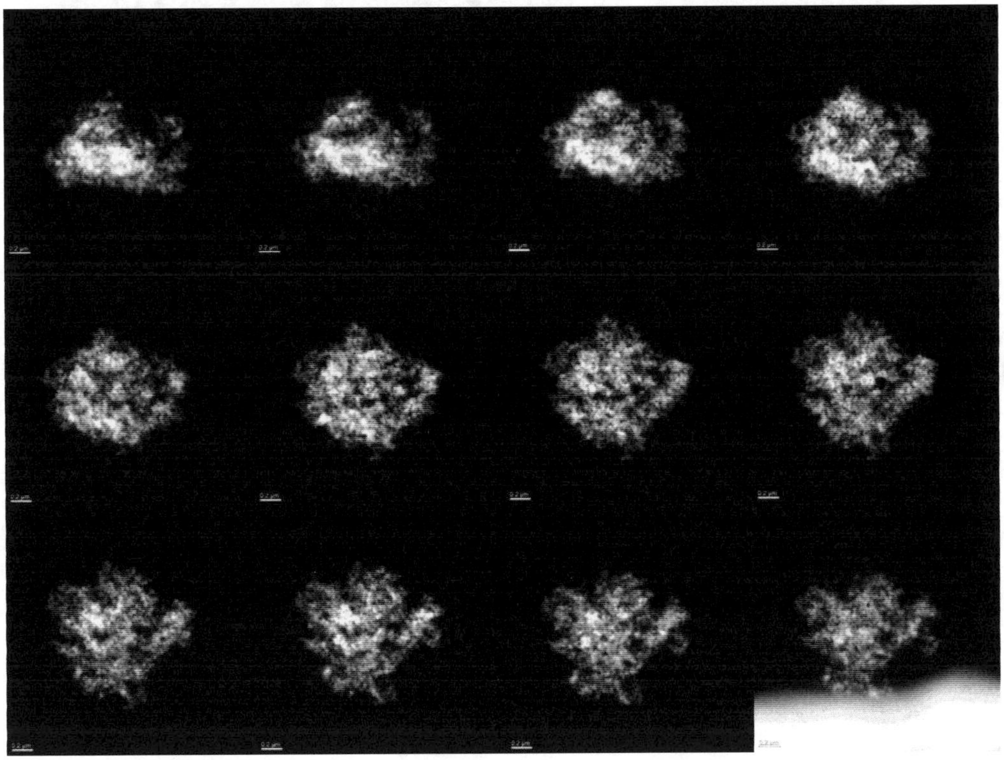

Figure 3. High angle annular dark field tilt series of a cluster of TiO_2 powder. Top left +60 degrees, bottom left -50 degree tilt, 10degree step between images. Bar is 200nm.

In Figure 4 are shown high resolution bright field images of Au nanoparticles on carbon film acquired at -60 degrees, zero and +60 degrees respectively. Information limit on the FFT corresponds to 1.06Å, indicating a good mechanical stability of the holder. This stability holds over time, in fact it has been possible to perform complete tilt series acquisitions every 2degrees from -60 to +60 on FIB samples (data not shown here). The 3D reconstruction will be published elsewhere.

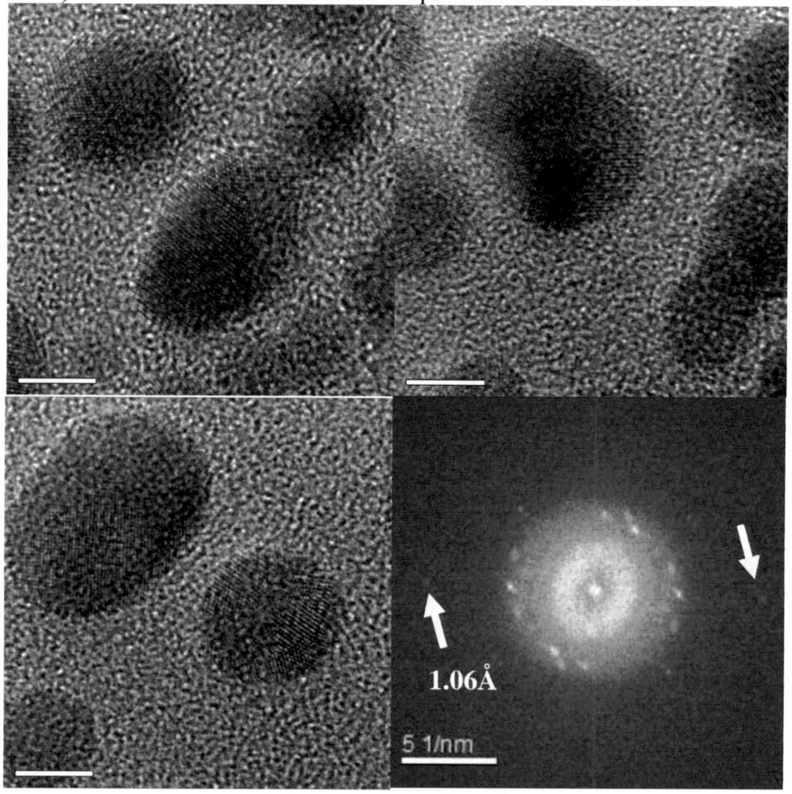

Figure 4. TEM HREM images of Au on carbon at **(Top left)** 60 degree tilt, **(Top right)** -60 degree tilt; **(Bottom left)** 0 degree tilt and **(Bottom right)** associated FFT.

4. Conclusions

In this paper we have shown that with a small investment a tomography capability can be implemented on a TEM/STEM microscope with a moderate gap pole piece. The high performance system shows mechanical stability and can receive standard 3mm TEM grids as well as FIB specimens. A simulation chamber replicating the pole-piece gap in the microscope was used to pre-test the holder to verify the clearance at maximum tilt. This development shows the versatility of the double aberration corrected York 2200FS TEM/STEM that, being equipped with a large gap pole piece, allows the use of special holders that would not fit the small gap of high resolution-pole pieces. The detrimental effects on point resolution of the large gap pole piece are compensated by aberration correction retaining 1Å resolution in vacuum.

Acknowledgments

We thank the EPSRC (UK) for research grant EP/J018058/1

References

[1] Hart R G 1968 *Science* **159** 1464
[2] Midgley P A and Dunin-Borkowski R 2009 *Nature Materials* **8** 271
[3] Messaoudi C, Boudier T, Sanchez Sorzano C O, Marco S 2007 *BMC Bioinformatics* **8** 288
[4] Gai P L and Boyes E D 2009 *Microscopy Research and Technique* **72** 153
[5] Boyes E D, Ward M R, Lari L and Gai P L 2013 *Annalen der Physik* **525** 423
[6] Arslan I, Tong J R and Midgley P A 2006 *Ultramicroscopy* **106** 994

Development of a cryo-FIB technique for the 3D structural characterization of liquid samples

M Tsuchiya[1], T Iwahori[2], A Morikawa[3], T Sato[3], Y Nagakubo[3]
1Hitachi High-technologies Europe GmbH, Europark Fichtenhain A 12, 47807 Krefeld
2 Hitachi High-Tech Science Corp., 36-1, Takenoshita, Oyama-cho, Shizuoka, 410-1393 japan
3 Hitachi High-Technologies Corp., 1040 Ichige, Hitachinaka, Ibaraki, 312-0033 Japan

E-mail: miki.tsuchiya@hht-eu.com

Abstract. The observation of three-dimensional distributions in some liquids or colloids is important, for example in cosmetics, functional paints and a catalyst. Furthermore, there is increasing demand to microscopically investigate the structures inside a liquid colloid, such as the interface of a dispersoid and dispersant. In order to meet these requirements, we have developed a cryo-transfer holder which is fully compatible between FIB and TEM/STEM. The cross section of frozen emulsions and a thin film sample were made by FIB-SEM (NB5000). As a result, the dispersion state of a dispersoid was observed by sequential FIB and SEM. Furthermore a 100nm thick sample was made by FIB and observed by 200 kV STEM (HD-2700). The dispersoid (TiO_2 and Fe_2O_3) morphology and distribution were shown at nanometer level.

1. Introduction

The focused ion beam (FIB) method is a well-established and useful technique for the preparation of cross sectional samples for SEM or TEM thin foils from solid materials such as semiconductor, ceramics, metals, and so on. But in recent years the FIB application field has been extended into macromolecules and biological materials. Due to the usually low thermal conductivity of such materials, the heat generated by ion beam causes a rise in sample temperature which eventually might prove fatal for the sample [1]-[2]. Thus, the demand for cryogenic sample cooling during FIB processing and subsequent SEM or TEM observation has increased [3] - [5].

In order to respond to those demands, we have developed a cryo-transfer holder (Fig.1) for Hitachi FIB and FIB-SEM which enables processing of temperature sensitive samples without heat damage. With this holder cryogenic sample applications can be directly executed in all Hitachi FIB and FIB-SEM systems equipped with the TEM compatible goniometer stage and on Hitachi TEM/STEM instruments, without any further modification required to the instrument. In this paper we report an application example for three-dimensional structure observation of a liquid sample by FIB-tomography using this cryo-transfer holder, and subsequent high-resolution observation in STEM

[1] Corresponding author Tel.:+49-(0)2151-6435-332; fax: +49-(0)2151-643564-332. Email address: miki.tsuchiya@hht-eu.com (Miki Tsuchiya)

2. Materials and methods

2.1. Thermometry of sample temperature using thermocouple holder
Two types of cryo-transfer holder were developed (Fig.1). One of them is a cryo-transfer holder for "bulk" specimens, and it is fully compatible with FIB and SEM. The other is for thin film specimens and is fully compatible holder with FIB, SEM and STEM or TEM systems.

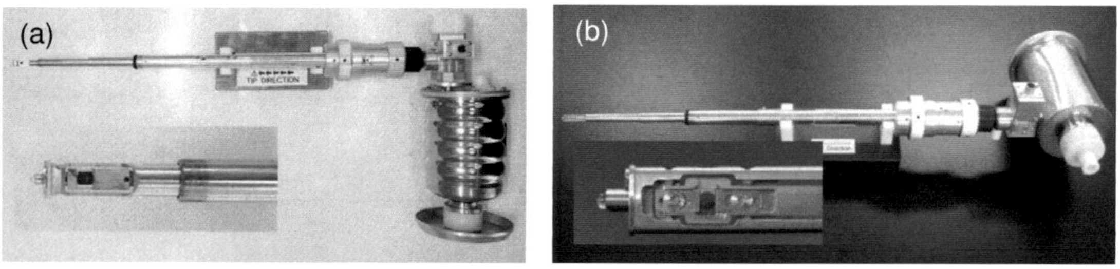

Figure 1. General view of the cryo-transfer holders. (a) For bulk sample. (b) For thin film sample.

To evaluate the cooling effectiveness of the cryo-transfer holder, we measured the temperature change of the sample surface under ion beam irradiation. The basis of the experimental instrumentation is the FB2200 FIB, together with a thermocouple holder (Hitachi, Japan) [6]. In the sample thermometry, the thermocouple holder can put into the micro-sampling stage of FB2200 and the thermocouple can be manipulated to directly contact the sample surface. We used copper as a sample for thermometry. Figure 2 shows the result of the thermometry. In the case of cryo-transfer holder for bulk specimen, the sample temperature fell to a minimum of -138.8 °C without ion beam irradiation. When applying a 40 kV ion beam with 64 nA beam current, the temperature change was small. In the case of the cryo-transfer holder for film specimens, the sample temperature fell to a minimum temperature -149.8 °C without ion beam irradiation. When applying a 40 kV ion beam with 64 nA beam current, the temperature increased to -147.8 °C.

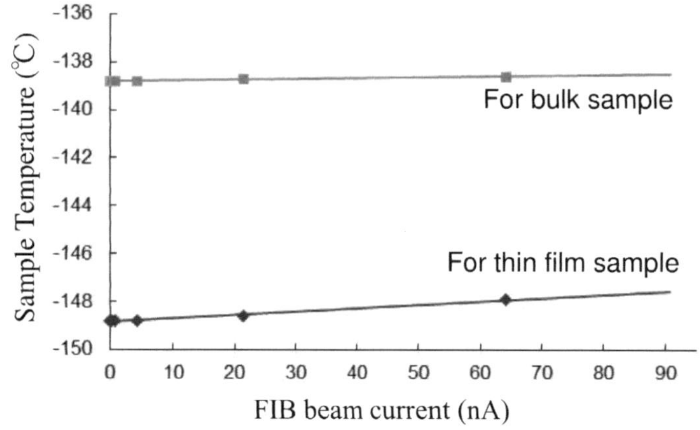

Figure 2. Thermometry of two types of cryo-transfer holder.
Sample: Cu
Holder Temperature:
Thin film: -170°C
Bulk: -149°C
FIB Vacc. :40 kV
FIB irradiation area: 37 μm×25 μm

2.2 Application of the cryo-transfer holder to a liquid sample
As a sample for our experiment we used a commercially available emulsion. This emulsion included mainly water, cyclomethicone, ethanol, glycerin and some metal oxides. Figure 3 shows the process of inserting the sample into the FIB-SEM for each cryo-transfer holder. Two experiments were performed. Firstly, cross sectional SEM observation of a "bulk" specimen and secondly, high resolution STEM observation of a thin film. In the first experiment, the sample was applied to a hydrophilized Cu stub / grid. After the sample was frozen by rapid freezing (ethane plunge), it was mounted on the pre-cooled cryo-transfer holder. The cryo-transfer holder was inserted into the

NB5000 FIB-SEM under constant cooling to -140 °C. During holder transportation the sample area was protected from air contact and frost formation by a retractable shielding capsule. We then applied the focused ion beam to produce a cross-section of the emulsion by using this cryo-transfer holder.
In the second experiment, we milled part of the sample in the Cu grid thinly and transferred it to HD-2700 200 kV STEM (Hitachi, Japan) for high resolution observation.

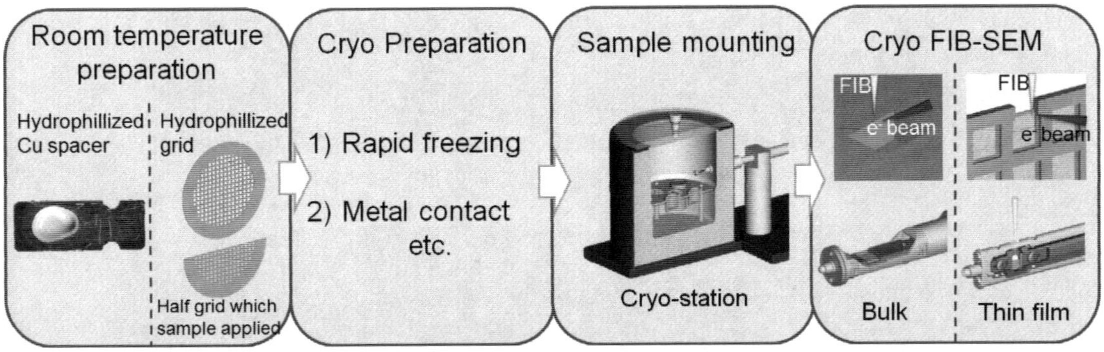

Figure 3. The procedure of the cryo-FIB thinning and STEM observation.

3. Results

Figure 4. shows a cross-sectional SEM image of the frozen emulsion. We can distinguish the dispersant (yellow arrow) from dispersoid (white arrow) in this image. We also observe that the surfaces of the dispersoid particles are coated with fine particles. Next, we obtained a series of 60 consecutive SEM images by using the automatic FIB milling and SEM observation cycle function of the NB5000. Figure 5 is a reconstructed 3D image of emulsion based on the serial section SEM images.

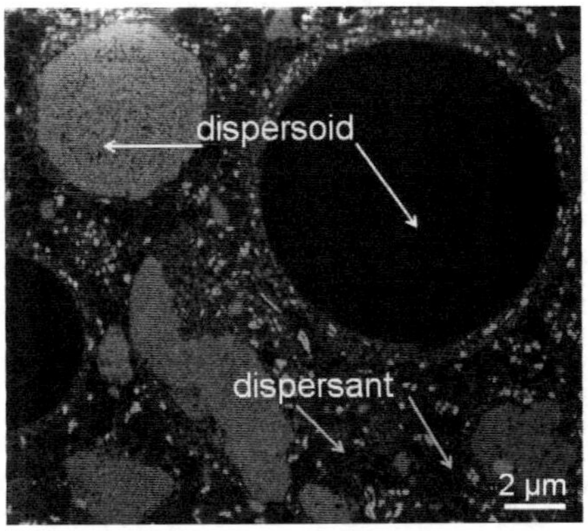

Figure 4. Cross-section image of emulsion. SEM Acc. Volt: 2.5 kV, Signal: Back scattered electron, Holder temp.: -140℃

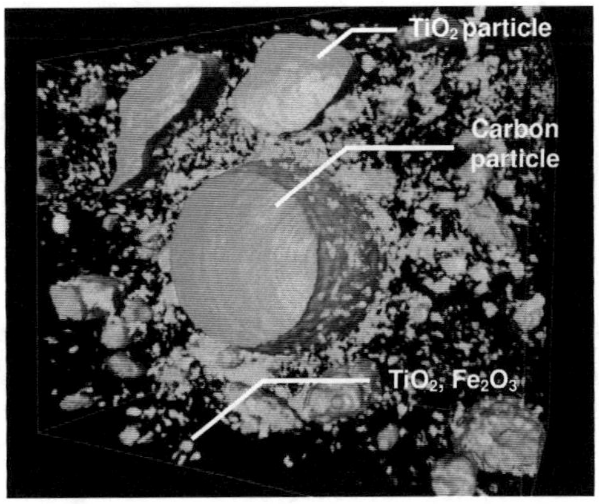

Figure 5. 3D reconstruction image.

The reconstruction volume is approximately 20 μm in width, 17 μm in height and 12 μm in depth. We were able to clearly visualize the three-dimensional distribution of the pigments in the emulsion.

Figure 6 shows the 200 kV STEM image of the thin film specimen (Thickness: approximately 100 nm) of the frozen emulsion. In this result it is possible to clearly see each dispersoid, such as the needle crystals and the granular crystals. The results indicate the low temperature and high stability performance of this cryo-transfer holder.

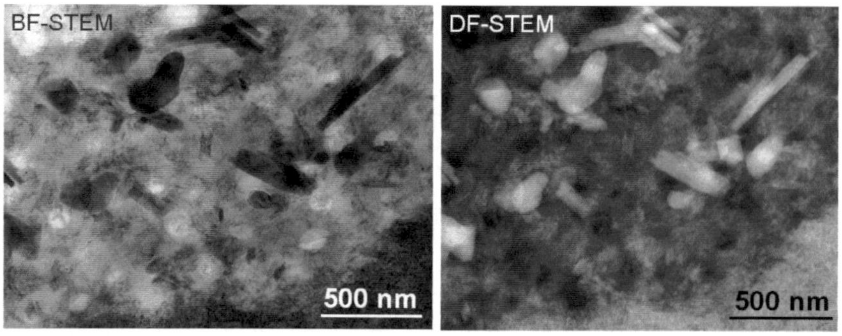

Figure 6. BF and DF-STEM images of emulsion.

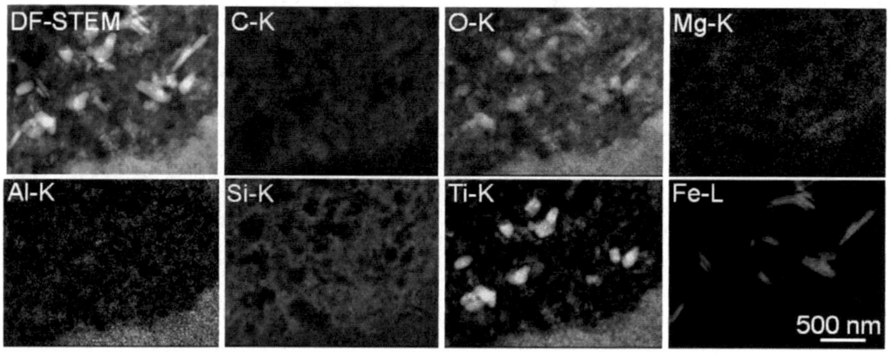

Figure 7. DF-STEM image and EDX maps of emulsion.

Figure 7 shows the EDX maps of emulsion, the needle crystals and granular crystals each represent of Fe and Ti. Very fine crystals also are composed of Ti and most of dispersion medium is made up of Si and carbon materials.

4. Summary

We have developed two types of cryo-transfer holder for FIB-SEM and validated their functionality by applying it to the FIB processing and SEM observation of a rapid frozen emulsion. We succeeded in visualizing the 3-dimentional distribution of the dispersoid using automated serial FIB milling and SEM observation, showing the effectiveness of the sample cooling. In addition, we carried out thinning by FIB and we transferred the thin section to 200 kV STEM and performed STEM observation. The dispersoid morphology and distribution were clearly observed and we were able to undertake detailed studies of the structures and composition. Also, the demand for microscopic examination of interfaces in emulsions is expected to increase in many fields including biomaterials, and a capability and workflow to do that has been demonstrated.

5. References

[1] Kudo M, Morikawa A, Sato T, Iwahori T, Nagakubo Y 2012 *Proc. Of the 68th annual meeting of the JSM* (Tsukuba: Japan) p 112

[2] Tsuchiya M, Iwahori T, Morikawa A, Konno M, Nagakubo Y 2013 *Proc. Of the 69th annual meeting of the JSM* (Osaka: Japan)

[3] Kupsta M, Li P, Elliott G, Mikula R and Malac M 2012 *Microsc.Microanal.* **18** (2) 656

[4] Wang Q, Uzunoglu E, Libera M and Chou T 2012 *Microsc.Microanal.* **18** (2) 626

[5] Rigort A, Bäuerlein F, Leis A, Gruska M, Hoffmann C, Laugks T, Böhm U, Eibauer M, Gnaegi H, Baumeister W and Plitzko J 2010 *J.Struct. Biol.* **172** 169

[6] Nippon Steel Corp., Hitachi Ltd., Sugiyama M, Okada M and Koike H, "Charged particle beam system having thermometric function" Japan Patent 3969935. 2007-06-15

Correlative Microscopy Techniques for the Analysis of Particles in Safeguards Environmental Samples

N Dzigal[12], **E Chinea-Cano**[1]
[1]International Atomic Energy Agency, Wagramerstrasse 1, 1220 Vienna, Austria
[2]Vienna University of Technology, Karlsplatz 13, 1040 Vienna, Austria

E-mail: n.dzigal@iaea.org, e.chinea-cano@iaea.org

Abstract. This paper presents a novel approach to environmental particle analysis for safeguards by means of a combination of micro-analytical techniques. It includes the tandem utilization of two separate light microscopes, a scanning electron microscope and a femtosecond laser-ablation ICP-MS. These are: a light microscopy automated particle relocation device (Zeiss Z2m); an optical-microscopy-based laser micro-dissection system (IX83 MMI+Olympus); a focussed ion beam scanning electron microscope equipped with a time-of-flight mass spectrometer extension (Tescan Lyra3) and a fs LA-ICP-MS (J200 from Applied Spectra Inc. and Thermofisher Scientific iCap Q). The samples examined in this contribution are analysed for their nuclear material signatures, in particular the presence of uranium isotopes.

Keywords: Correlative microscopy, Safeguards, Laser micro-dissection, LA-ICP-MS, FIB-SEM-TOF, Uranium isotopes,

1. Introduction

Environmental samples are processed and analysed in the Environmental Sample Laboratory (ESL), which is part of the Safeguards Analytical Laboratories (SGAS) of the International Atomic Energy Agency, Seibersdorf, Austria. The samples are exclusively handled in cleanroom facilities where all measurements are conducted using micro- and ultramicro-analytical techniques. Advanced instruments such as the LG-SIMS, TIMS, ICP-MS, LA-ICP-MS and SEM are routinely used alongside established radiochemical methods for the bulk and particle analysis of uranium-containing swipe material. [1] [2]

Correlative microscopy has a growing number of applications in the biological [3] [4] [5] [6] [7] and materials [8] [9] [10] sciences. The basic idea is to define a site of interest and then relocate the same area in another instrument for further sample characterization. This paper describes recent efforts in the relocation of 1-5 μm particles from a single sample between two optical microscopes, a scanning electron microscope and a LA-ICP-MS instrument. In addition to correlative microscopy, this contribution goes beyond and demonstrates a method whereby sites of interest were physically isolated and transferred onto a separate substrate for further microprobe analyses steps. In the field of Safeguards, no single microscopy or mass spectrometry technique can provide a complete assessment of an environmental swipe sample (ESS) which is why a particle tracing scheme is needed for facilitating combined analysis from several techniques.

[1] To whom any correspondence should be addressed.

An analytical procedure was developed for the systematic processing of ESS that includes (Figure 1.):

1. Deposition of sample particles from a swipe onto a catcher;
2. Identification of Particles-Of-Interest (POIs) on the catcher;
3. Isolation and transfer of POIs onto a harvester;
4. Microprobe analysis of POIs;

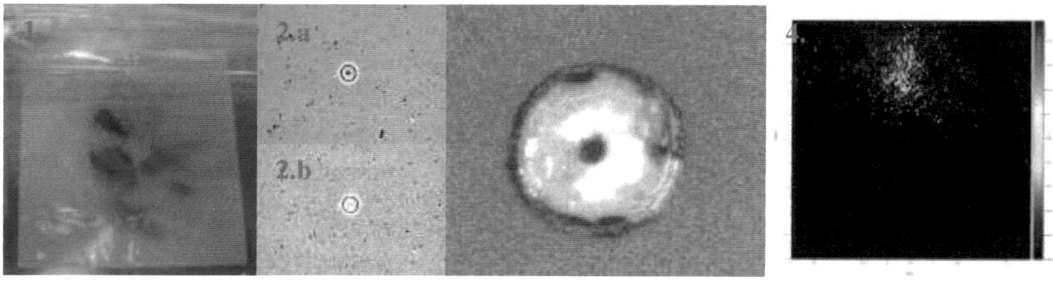

Figure 1: Analytical procedure steps: 1. Collection of particles from swipe; 2.a) Dispersion and laser microdissection of POIs; 2.b) isolation of POIs 3. transfer of POIs onto a harvester; 4. Microprobe analysis of POIs (FIB-SEM-TOF U-238 m/q signal is depicted.)

2. Sample Preparation for Microprobe Analysis

Figure 2: Laser markings for relocation. The fiducial points are: on the triangle, the end of the line bisecting the upper angle; square - centre of cross; circle - centre of cross.

2.1. Deposition of particles onto a catcher (dispersion)
MMI membrane slides with stainless steel rims were used as the catcher substrate. These were marked with three fiducial points to be used for later relocation of POIs [3]. The markings are shown in Figure 2. and covered an area of 10mm x 17 mm. The slide was coated with clear varnish (Daler-Rowney Ltd, Bracknell, Berkshire, USA) by means of a spin coater. After this, the sample was placed inside a shock-wave disperser for deposition of particles from the swipe onto the varnish-coated membrane slide. The test material was a homogenized and sieved DU contaminated soil. The loading of material on the shock wave disperser was ~ 40 μg. Details of the dispersion method are discussed elsewhere [4].

2.2. POI identification.
After dispersion, the sample was allowed to settle for 24 hours before coating with Au. Six particles containing DU were identified by a TESCAN Lyra3 SEM. Figure 3. displays an EDX spectra from a single DU-particle (X Max-50, Oxford Instruments). The coordinates of the three reference points and six POIs was recorded for relocation in other instruments.

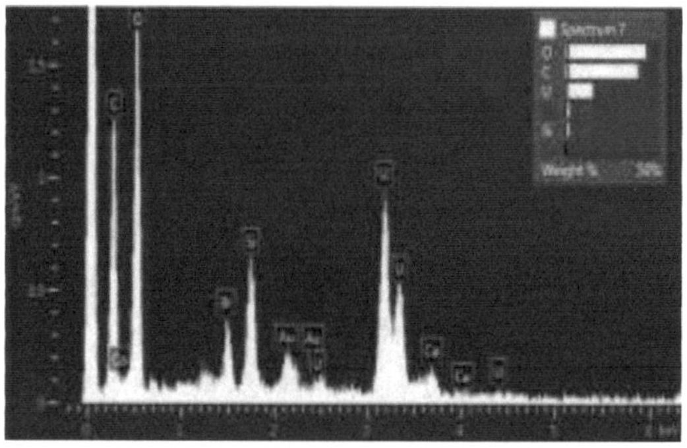

Figure 3: EDX spectra of a single DU particle

2.3. Relocation of POIs
An overview image of the fiducial points and POIs was made using a Zeiss Z2m light microscope along with its accompanying software (see Figure 4.) About 80 individual images taken at 5x magnification (1.268 μm/pixel resolution) were used to build the fused sample collector image. The objective used was an EC Epiplan-Neofluar 5x/0.13 HD M27 with a working distance of 15.1 mm. The images were stitched by the native AxioVision SE64 software using stage coordinates and an image overlap of 10%. No filtering was conducted to enhance

Journal of Physics: Conference Series **644** (2015) 012015 doi:10.1088/1742-6596/644/1/012015

image features and the images were only corrected for background illumination during the acquisition (flat-field correction). The relocation accuracy was ~ 20μm.

2.3.1. Laser micro-dissection of POIs.

The laser microdissection device (LMD) used was an MMI IX83 with Olympus scope. A 3-point-based algorithm *[11]* for the transfer of fiducial points and POIs was used to generate a POI relocation file with MMI stage coordinates. The POIs were found in the LMD device with an accuracy of ~ 50μm.

A harvester was made out of MMI microdissection chambers (diameter 35 mm) with a silicone bottom. The transparent silicone was removed and trimmed to fit the slit on the membrane slide and carefully fixed onto a standard microscopy glass slide (76mm x 26mm). The harvesting was done by placing the catcher on top of the harvester so that the silicone was under the membrane slide and the deposited particles were on top, see Figure 5.) This prevented unwanted particle transfer to the harvester, see Figure 6. for laser micro-dissection illustration). A 50 μm circle was cut around each particle (see Figure 8. for an image of the catcher after LMD cuts and detachment of the harvester). The cut parameters were: cut velocity 27 μm/s, laser focus 2667 μm and 80% laser energy. During the laser-cutting, the membrane with the POIs would stick onto the melting silicone. New markings were made around the POIs for facilitating relocation in other instruments (see Figure 7. for an SEM image of a relocated DU particle and Figure 8. for an overview of the catcher post cutting). Additional material including videos of the dissection can be found online.

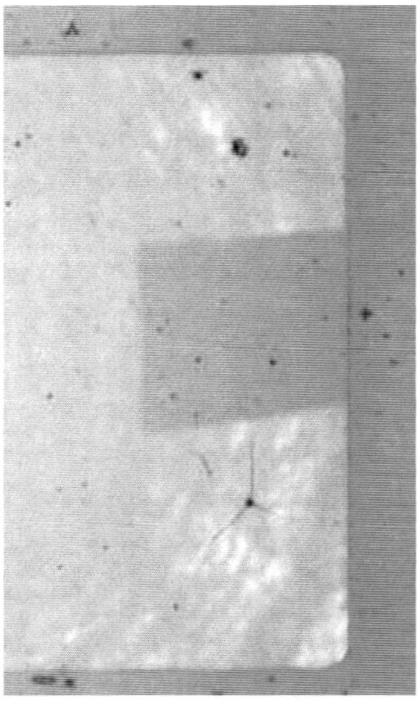

Figure 4: Overview of relocation area. Laser markings are shown: circle (bottom left, diameter 117 um), square (right, 190 um side length) and triangle (tope left, base length

Figure 5: (top) harvester with silicone; (bottom) MMI membrane slide was placed on top of the harvester.

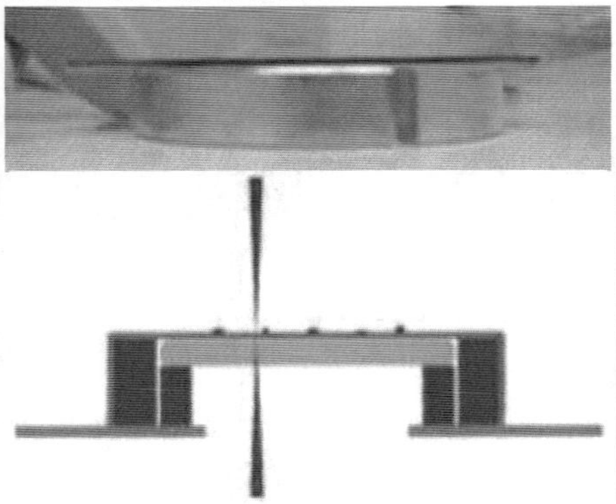

Figure 6: (top) side-view of catcher and harvester sandwich; (bottom) laser-microdissection of POI.

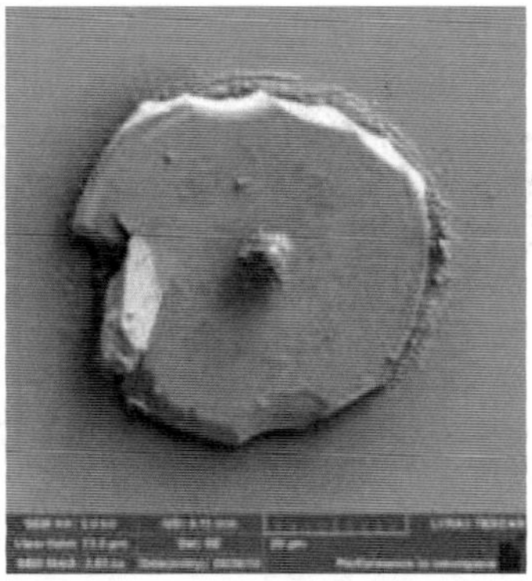

Figure 7: A relocated POI with membrane material. The image was taken in a Lyra3 SEM prior to FIB-SEM-TOF analysis.

3. Results of Microprobe Analysis

Prior to transfer into an FIB- SEM-TOF, the harvester and catcher were detached. All particles were transferred successfully onto the harvester. The TOF measurement was conducted in a Lyra3 FIB-SEM-TOF instrument – see Figure 1 Part 4 for a laterally resolved signal of a single DU particle (the U-238 mass spectra signal is shown). The uranium isotopic composition of a 1μm particle obtained from the FIB-SEM-TOF was evaluated at 0.0046 (U-235/U-238).

The harvester with the remaining five POIs was transferred into a J200 LA/LIBS system (Applied Spectra Inc.). These particles were ablated with the following parameters: 3 shots @ 30 μm spot size, 40 Hz repetition rate, 100% output energy, 0.7 l/min Ar flow and 0.7 l/min He flow. The J200 LA system was connected to a quadrupole mass spectrometer - iCap Q (Thermofisher Scientific) to obtain simultaneous mass spectrometry measurements of the five POIs. It was found that the mean isotopic ratio of uranium for the five particles was 0.0020 U-235/U-238 (naturally occurring U is found in the isotope ratios: 0.0072 U-235/U-238) *[12]*. See Figure 9 for an overview image of the harvester post laser-ablation.

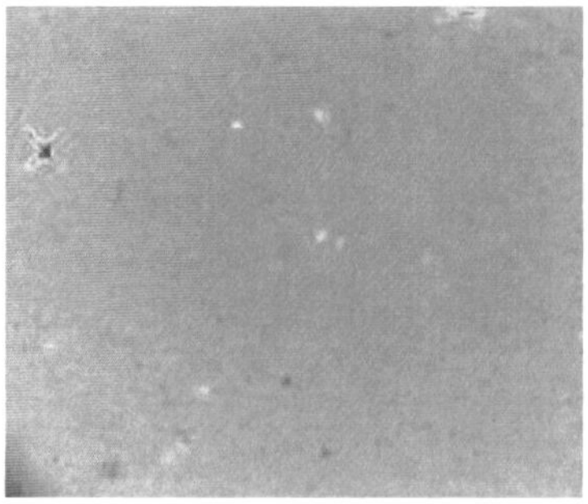

Figure 8: Catcher without POIs and new relocation points.

4. Conclusion

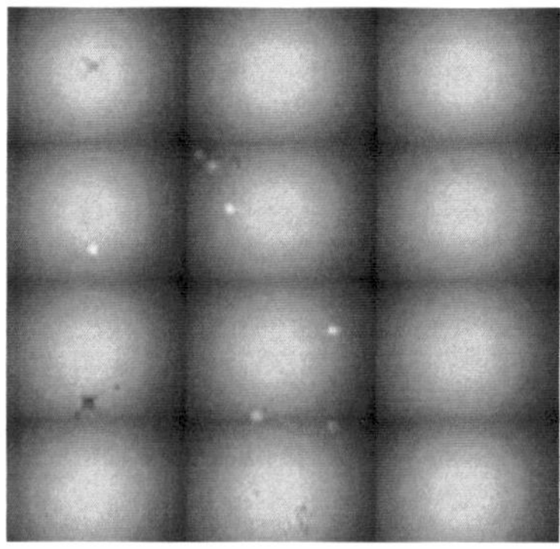

Figure 9: An overview image post laser-ablation in the J200 system. Note the relocation points circle (upper left); square (lower left) and triangle (lower right). The acquisition of TOF data on a single particle was conducted prior to LA-ICP-MS measurements. One spot (top most of ablation points) was used for the testing of the laser-ablation signal and background measurement of the silicone substrate.

The proposed analytical procedure for the isolation and transfer of six 1-5 μm particles was found to be successful. A mixture of soil particles with depleted uranium were deposited onto a thin membrane slide and six particles containing DU were identified by means of their EDX spectra in an SEM. Their positions were recorded with reference to three fiducial marks for later relocation in a LMD device. Once transferred, they were cut out and collected onto a silicone-based harvester. All particles were harvested successfully. One was analysed by means of a FIB-SEM-TOF and five by LA-ICP-MS. All micro-probed particle measurements found the U isotopic signature of the particles to be of depleted uranium (i.e. less than 0.0072 U-235/U-238).

5. References

[1] D. Donohue, *Analytical Chemistry,* vol. 74, no. 1, pp. 28A-35A, 2002.

[2] G. Voigt and M. Scheland, IAEA Internal SGAS Report, Vienna, Austria, 2013.

[3] A. Sartori, R. Gatz, F. Beck, et al, *Journal of Structural Biology,* vol. 160, no. 2, pp. 135-145, 2007.

[4] J. Plitzko and A. L. A. Rigort, *Current Opinion in Biotechnology,* vol. 20, no. 1, pp. 83-89, 2009.

[5] P. Kempen, M. Kircher, A. de la Zerda, et al, *Micron,* vol. 68, pp. 70-76, 2015.

[6] C. Fonta and B. Humbel, *Archives of Biochemistry and Biophysics,* 2015.

[7] D. Kong and J. Loncarek, *Methods in Cell Biology,* p. online, 2015.

[8] Carl Zeiss, *Tools & Techniques,* vol. 12, no. 10, p. 47, 2009.

[9] A. Fejfar, M. Hyvl, A. Vetuschka, P. Pikna, et al, *Solar Energy Materials and Solar Cells,* vol. 135, no. EMRS 2014 Spring Meeting – Advanced materials and characterization techniques for solar cells II, pp. 106-112, 2015.

[10] M. Herbig, P. Choi and D. Raabe, *Ultramicroscopy,* vol. 153, pp. 32-39, 2015.

[11] U. Admon, et al., *Hot Particles Released from Different Nuclear Sources*, Yalta, Crimea, 2007.

[12] S. Richer, A. Alonso, et al, *International Journal of Mass Spectrometry,* vol. 193, no. 1, pp. 9-14, 1999.

[13] U. Admon, E. Boblil, E. Chinea-Cano and N. Dzigal, *IAEA Safeguards Symposium*, Vienna, 2014.

6. Acknowledgements

The authors thank Stefanie Konegger-Kappel (IAEA), Jhanis Gonzalez (Applied Spectra Inc., Lawrence Berkeley National Laboratory) for LA-ICP-MS measurements, Libor Sedláček and Kateřina Klosová's assistance with FIB-SEM-TOF spectra (both from Tescan, s.r.o).

Achieving 50 nm lateral-resolution quantitative EDX SEM

G Pimentel[1], S Lozano-Perez[1]

Department of Materials, University of Oxford, Oxford OX1 3PH, UK

E-mail: gemma.pimentel@materials.ox.ac.uk

Abstract. Low Voltage Scanning Electron Microscopy (LV-SEM) has become a very promising approach to perform Energy Dispersive X-ray (EDX) chemical mapping with high-lateral resolution [1]. Using voltages as low as 1.5keV, sub-10nm resolutions can be achieved. In this work, we try to take advantage of the small interaction volume in order to simplify the otherwise more complex SEM quantitative methodology. This way, phenomena such as absorption and fluorescence can be ignored and, effectively treat the quantification as with the Transmission Electron Microscopy (TEM)-based Cliff-Lorimer method. Experimental k-factors have been obtained from a series of standards and used to quantify complex oxide phases in steels.

1. Introduction

Scanning Electron Microscopy (SEM) quantitative X-ray microanalysis of bulk samples has been traditionally performed at 20 – 30 keV by integrating X-ray peaks intensities and comparing them to those of relevant standards. The role of the interaction volume at these high voltages is taking into account by correcting for atomic number, absorption and fluorescence (ZAF) effects. In order to simplify the calculation, the ratio of the extracted intensities to that of the standard is related to the ratio of the concentrations by using the ZAF or the $\varphi(\rho z)$ methods [2]. When using voltages below 5keV, the volume where electrons can produce characteristic X-lines will be smaller than 100 nm for most element of interest and, therefore, the ZAF corrections might be neglected, i.e. the SEM sample could be treated as a "thin" sample, and the X-ray quantification could be calculated with the Cliff-Lorimer approach. In this study, we will demonstrate that this is the case for surface oxides in stainless steels. Oxides are traditionally challenging to quantify by Energy Dispersive X-ray (EDX), but the right quantification calibration can make this a very easy-to-use and reliable method.

2. Materials and experimental method

We will use commercially available stoichiometric oxides and various metallic alloys which composition was validated by gas chromatography for the calculation of experimental k-factors. Their compositions are shown in Table 1. Then, we will use them to verify that the composition of in-service oxides is the same that was calculated by Transmission Electron Microscopy-Electron Energy Loss Spectroscopy (TEM-EELS) and/or Atom Probe Tomography (APT).

X-Ray spectra were collected with a Zeiss Merlin Field Emission Gun (FEG)-SEM equipped with an Oxford Instruments detector (X-Max 150mm^2). Acquisitions were made at 1.5, 2, 3 and 5 keV; with an acquisition time of 30 s, the slowest processing time and a dispersion of 5eV/channel.

Peak area integration was performed via background and peak deconvolution, using NIST's DTSA-II software [3]. The k-factors were then calculated for each nominal composition with the Cliff-Lorimer equation [2] using K lines for O and L lines for Fe, Cr and Ni.

Table 1. Wt. % composition of standards and reference alloys.

	Wt. %				
	O	Fe	Cr	Ni	Other Elements
FeO	22.3 ± 0.5	77.7 ± 0.5	-	-	-
Fe_2O_3	30.1 ± 0.5	69.9 ± 0.5	-	-	-
Fe_3O_4	27.6 ± 0.5	72.4 ± 0.5	-	-	-
304 SS	-	70.0 ± 0.5	18.3 ± 0.5	9.2 ± 0.5	2.5 ± 0.5
Alloy 600	-	8.39 ± 0.05	16.07 ± 0.05	73.8 ± 0.05	1.74 ± 0.05

3. Results and discussion

3.1. Interaction volume at low High Tension (HT)

One of the benefits of working at low HT (< 5kV) is the smaller interaction volume reached at these voltages. The interaction volumes at different HT were simulated for the three standard oxides and reference alloys using Casino software [4]. Figure 1 represents the lateral and depth dimensions from which 95% of the Fe L X-rays were generated for standards and references at each HT. It can be seen that the interaction volume and spatial resolution is dependent on the element composition and below 100 nm in most cases.

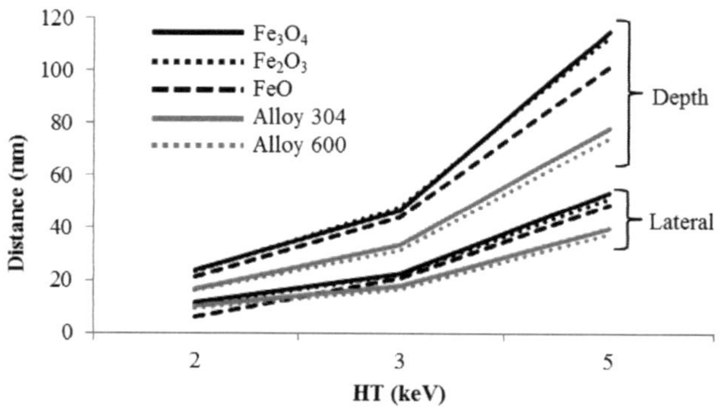

Figure 1. Achievable lateral and depth resolutions (95% of the X-ray emitted) for the oxide standards and reference alloys at different HT calculated from the simulations.

3.2. k-factor determination

Experimental k-factors for Fe and O were calculated from oxide standards (k_{Fe-O}) at 5, 3, 2 and 1.5 keV. For Fe, Ni and Cr, k-factors from reference alloys (k_{Fe-Ni}, k_{Fe-Cr}, k_{Ni-Cr}) were calculated at 2 keV since this is the voltage used for the rest of the results presented in this paper (as shown in Table 2).

Table 2. Average experimental k-factors calculated from the standards and reference alloys.

	Average k-factor (exp)			
Volt. (kV)	$k_{O\text{-}Fe}$	$k_{Fe\text{-}Ni}$	$k_{Fe\text{-}Cr}$	$k_{Ni\text{-}Cr}$
5	0.88 ± 0.03			
3	0.86 ± 0.01			
2	0.83 ± 0.06	1.51 ± 0.04	0.90 ± 0.04	0.60 ± 0.02
1.5	0.74 ± 0.07			

The small relative error is promising for attempting the quantification of oxides and alloys. The lower 1.5kV value might indicate the voltage effect in the generation of Fe L counts.

3.3. Validating the simulations

Simulated spectra were generated with NIST-II software from all standards at a voltage range from 2 to 20 keV and k-factors extracted in a similar way. The software simulates the corresponding spectra from a given atomic composition (analytical simulation) modelling the atomic excitation process using a φ(ρz) curve, representing the efficiency of generation of atomic excitation as a function of the depth of penetration into the material.

The average k-factor values at every beam current for experimental and simulated data are represented in Figure 2, where a remarkable agreement is found. Their use would allow a shorter calibration routine.

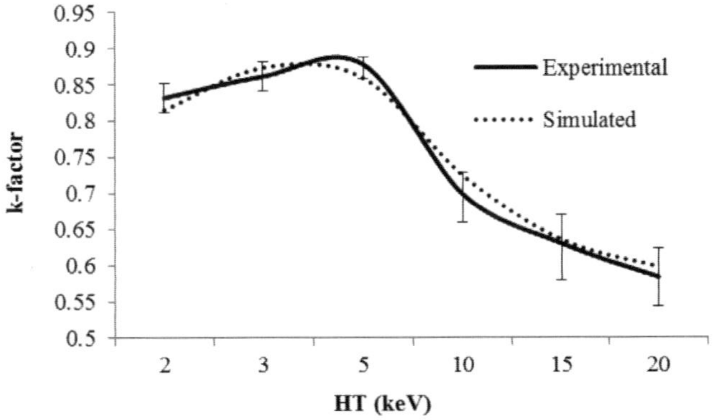

Figure 2. Experimental (solid line) and simulated (dotted line) average k-factors represented as a function of the HT.

3.4. Complex oxides quantification

Experimental k-factors were tested on two real oxides of "unknown" composition present in a 304SS alloy which was oxidized under simulated Pressurized Water Reactor (PWR) primary water conditions [5]. Figure 3a shows a cross-sectional area of 304SS alloy where spectrums 304SS_OX1 (Figure 3b) and 304SS_OX2 (Figure 3c) were collected from two different oxides next to the surface of the alloy. Both spectra show the C K line peak at 0.28 keV. Since this peak is only due to surface contamination, it has not been taken into account for the quantification. Also, a small Ni L peak at 0.85keV can be observed in the spectrum from region 304SS_OX1 and has not been taken into account in the final quantification, since it is believed to come from the neighbouring Ni-rich metallic regions surrounding the Cr-rich oxide [5]. EDX spectra from the two "unknown" surface oxides were quantified using the experimental k-factors calculated above, as shown in Table 3.

Table 3. Experimental and "expected" composition of the two oxides present in the 304 SS alloy.

	Element	At. calculated%	At. expected %
304SS_OX1	O	58 ± 1	59.0 ± 0.5
	Cr	23 ± 1	21.0 ± 0.7
	Fe	19 ± 1	20.0 ± 0.7
	Element	**Wt. calculated%**	**Wt. expected%**
304SS_OX2	Fe	77 ± 1	75.0 ± 0.5
	O	23 ± 1	25.0 ± 0.5

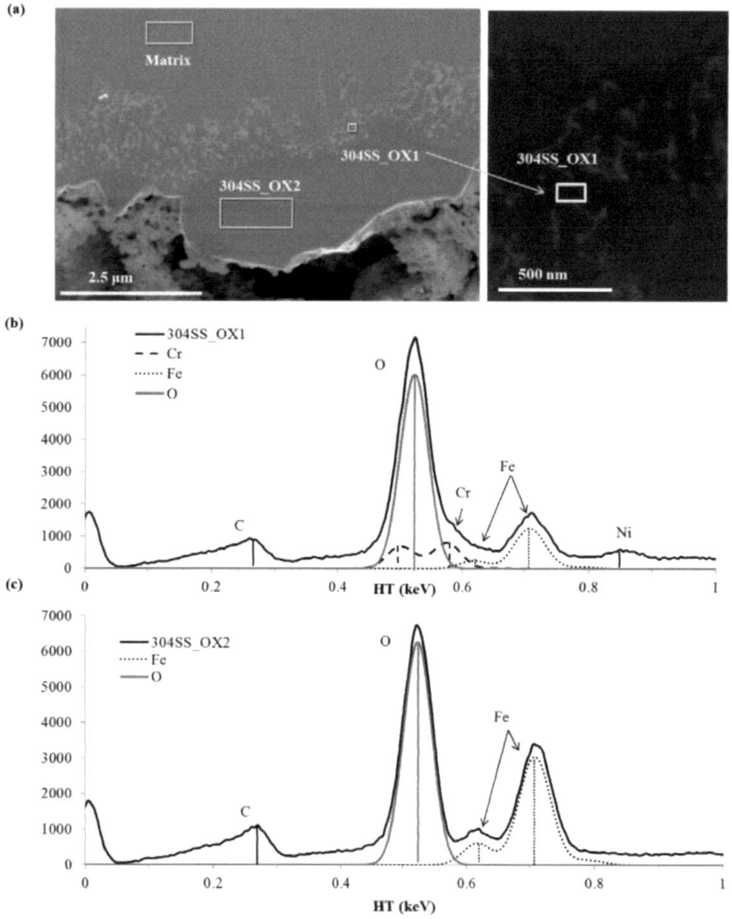

Figure 3. (a) Alloy 304SS SEM micrographs showing three areas where X-ray analysis was performed: matrix, oxide named 304SS_OX1 and oxide named 304SS_OX2. Experimental X-ray spectra (taken at 2 keV) and elemental deconvolution b) oxide 304SS_OX1 and c) oxide 304SS_OX2.

It can be said that the oxide of the spectrum 304SS_OX1 matches well with a $(FeCr)_3O_4$ –type spinel– and spectrum 304SS_OX2 with Fe_3O_4. In this regard, comparing calculated and nominal composition (also shown in Table 3) it is demonstrated the potential of this approach to quantify surface oxides with high-spatial resolution. It should be noted that oxides from the same samples, with the same morphology and microstructure have been characterized by TEM-EELS and APT in our group obtaining similar compositional results. The low-keV and its corresponding reduced interaction volume, allows the bulk samples to be treated as "thin" samples, in a similar way to TEM samples, which leads to the use of Cliff-Lorimer approach after careful experimental k-factor determination.

3.5. Surface contamination effect

Due to the relatively small interaction volume, surface contamination can become a limiting factor for a correct quantification/calibration. This effect will be higher at lower voltages and will result on different peak intensities ratios, i.e. different k-factor. In figure 4 it is represented the k-factor for the FeO as a function of the carbon thickness. It is clear that the k-factor is affected from C thickness thicker than 25 nm.

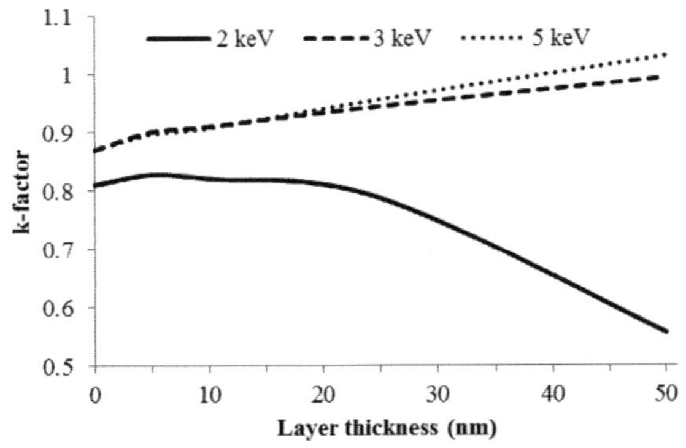

Figure 4. Simulated k-factors at 2, 3 and 5 keV represented as a function of C layer thickness.

In base of that, if the k-factor had been determined from a contaminated surface with a 50 nm thick C layer, the composition (wt.%) of the called 304SS_OX2 would change to 83Fe17O instead of the 77Fe23O.

4. Conclusions

The feasibility of using the Cliff-Lorimer approach for the X-rays quantification of SEM samples at low beam currents has been studied and has led to the following conclusions:

- Analytical resolutions smaller than 50 nm can be achieved with SEM-EDX.
- k-factors can be effectively "simulated", which simplifies the calibration procedure for alternative voltages.
- The small relative error in the experimental k-factors and the good compositional quantification match obtained when applied to "unknown" oxides make it a promising technique for the quantification of oxides and other low Z phases.
- Surface contamination can have an important effect on the k-factor determination (especially at HT < 3keV) and should be minimized when possible.

5. References

[1] Meisnar M, Lozano-Perez S, Moody M, Holland J, 2014 *Micron* **66** 16.
[2] Goldstein JI, Newbury DE, Joy DC, Lyman CE, Echlin P, Lifshin E, Sawyer L, Michael JR 2003 *Scanning Electron Microscopy and X-Ray Microanalysis*. New York: Plenum Press.
[3] Ritchie NWM, Davis J, Newbury DE, 2008 *Micros. Microanal.* **14** (SUPP 2) 1176.
[4] Drouin D, Real Couture A, Joly D, Tastet X, Aimez V, Gauvin R, 2007 *Scanning* **29**(3) 92.
[5] Lozano-Perez S, Kruska K, Iyengar I, Terachi T, Yamada T, 2012 *Corr. Sci.* **56** 78.

Acknowledgements

Authors are grateful to Oxford Instruments (S. Burgess and J. Holland) for support and to INSS (Japan) for sample provision.

Application of low-voltage backscattered electron imaging to the mapping of organic photovoltaic blend morphologies

R C Masters[1], Q Wan[1], Y Zhou[2], A M Sandu[3], M Dapor[4], H Zhang[2], D G Lidzey[5] and C Rodenburg[1]

[1]Department of Materials Science and Engineering, University of Sheffield, Sir Robert Hadfield Building, Mappin Street, Sheffield S1 3JD, UK
[2]School of Physics, Trinity College, Dublin 2, Ireland
[3]FEI Co. Europe NanoPort, Achtsewag Noord 5, Eindhoven 5651 GG, Netherlands
[4]ECT*-FBK and TIFPA-INFN, via Sommarive 18, Trento I-38123, Italy
[5]Department of Physics and Astronomy, University of Sheffield, Hicks Building, Hounsfield Road, Sheffield S3 7RH, UK

Email: rmasters1@sheffield.ac.uk

Abstract. With organic photovoltaic (OPV) technology moving towards commercialisation, high-throughput analytical techniques are required to study the nanoscale morphology of OPV blends. We demonstrate a low-voltage backscattered electron imaging technique in the SEM that combines a solid-state backscattered electron detector with stage biasing to produce a rapid overview of the phase-separated surface morphology of an organic photovoltaic (P3HT:PCBM) blend. Aspects of obtaining the best possible results from the technique are discussed along with the possibility of probing the sub-surface morphology by altering the primary electron beam landing energy.

1. Introduction

Organic photovoltaics (OPV), consisting of a polymer:fullerene blend, are a rapidly developing alternative to conventional photovoltaic technology [1]. Recent advancements in efficiency [2] are pushing OPV devices ever-closer to becoming a feasible option for wide-scale commercialization. It has long been accepted that the (hierarchical, and often nano-scale) morphology of the OPV blend plays a key role in defining the PCE of a given blend system [3]. In order to inform the development of more efficient and consistent devices in the future, it is therefore essential to provide fast, effective tools that can provide high-resolution morphology maps for analytical purposes.

Here we present a low-voltage imaging technique in the SEM using a solid-state backscattered electron (BSE) detector as a technique enabling rapid, easily acquired overviews of OPV morphology.

2. Experimental

We produced P3HT:PCBM blends films as described in [4]. In some cases, these films were subsequently thermally annealed (for 10 minutes at 150°C) in order to encourage the formation of a coarser phase structure more representative of those employed in actual OPV devices.

All imaging was performed using the solid state concentric backscatter (CBS) detector on a FEI Nova NanoSEM 450, at a working distance of 4 mm. The detector comprises an annular design (see

Figure 1a), split in to 4 segments of increasing radius labeled A-D. These segments can be used in any combination, in order to optimize the quality of the acquired images.

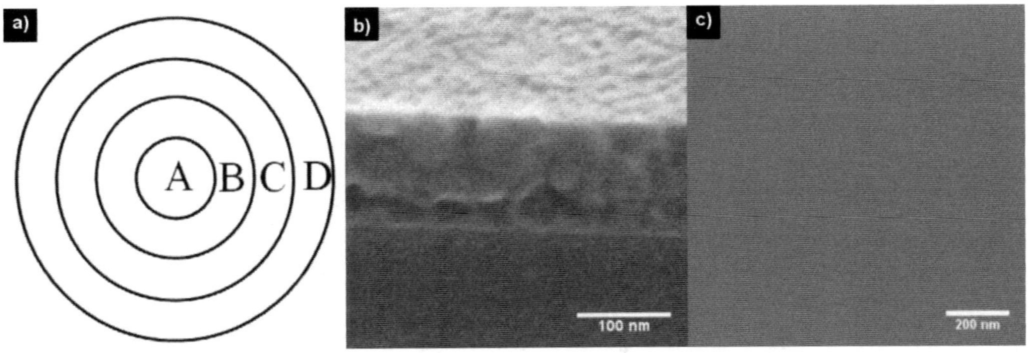

Figure 1. a) schematic of CBS detector design, b) Cross-section image of blend film as imaged in a Helium-ion microscope, used to measure blend film thickness, highlighted bright line represents the silicon-film interface. c) Example image of P3HT:PCBM blend film when using CBS detector without stage biasing.

Due to the film being relatively thin (~130 nm, as measured by helium-ion microscope cross section, Figure 1b), primary electrons with landing energies ranging from 500 eV to 3 keV were used to limit the interaction volume to the film as far as possible. Combining a solid-state BSE detector with low-energy primary beams, however, severely limits both detected signal strength (solid-state detectors have low detection efficiency below a few keV) and lateral spatial resolution in the resulting images (see Figure 1c) [5]. Applying stage biasing as a beam deceleration field has been shown to significantly improve image quality in this regard [5], by both increasing the number of electrons incident upon the CBS detector and improving the electron detection efficiency. Imaging resolution is also improved at low-kV as the electron beam is produced at a higher energy (giving inherently better resolution) before being decelerated to the desired landing energy. All images in this work have therefore used a -4kV specimen bias acting as a beam deceleration field.

3. Surface-level morphology data

To acquire surface-level morphology data, a beam landing energy of 500 eV was used in order to minimize the interaction volume of the beam. As an approximation, simplified models run in CASINO [6] suggest a maximum BSE interaction depth of ~8nm at 500 eV. Surface morphology images taken of P3HT:PCBM blend films are presented in Figure 2. We note that if a fresh blend film is imaged in this way, limited contrast is seen, which we have previously found to result from a P3HT-rich surface layer obscuring the blend morphology beneath [4]. This layer can be removed by a 5-minute low-power plasma etch in air using the FEI Nova's in-chamber procedure.

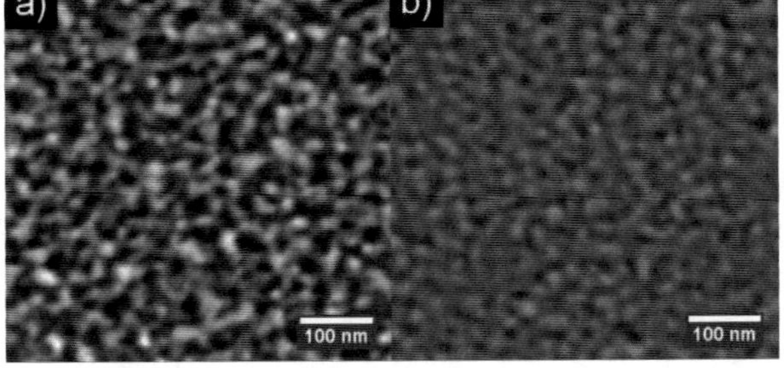

Figure 2: P3HT:PCBM blends imaged using CBS detector (segment A) at 500 eV landing energy. a) has been subject to a thermal anneal, b) is an as-cast film. The focus in both images here has been optimized; the lack of sharpness in b) is specific to the as-cast film.

We imaged P3HT:PCBM films both as-cast and after a thermal anneal. Clearly, the morphological difference between these two films can be observed in Fig. 2, with the thermally annealed sample displaying well-defined, clear phase separation and the as-cast film showing significantly reduced contrast between phases with less definition. We have measured the lateral resolution as 8.5nm in Figure 2a using the SMART-J plugin [7] for ImageJ, which we note is inferior to the current best-resolution techniques [5,6]. However this is to be expected – the use of BSE in this technique will limit the available resolution. Nonetheless, the results appear competitive with those acquired from bright-field TEM experiments [9] whilst negating the need for an electron-transparent specimen.

4. Effect of changing the primary beam landing energy

By increasing the primary beam landing energy, the interaction volume of the beam will increase, and BSE will originate from greater depths within the film. Imaging a blend film with a range of primary beam landing energies therefore allows the subsurface structure to be probed. Figure 3 shows the effect of increasing the beam energy from 500 eV to 3 keV on an as-cast P3HT:PCBM film - an increase from ~8 nm to ~130 nm in maximum BSE escape depth. A coarsening of the imaged morphology can be observed as the landing energy is increased. We interpret this as being representative of averaged contrast between the two phases projected over gradually increasing depths through the film, although some coarsening is due to the increased lateral extent of the interaction volume (up to 100 nm at 3keV, compared to ~10 nm at 500 V, according to CASINO simulations). This also affects the imaging resolution, which may explain the loss of image clarity at higher kV. We also note the generally reduced image quality in Figure 3 in comparison to Figure 2 – no plasma clean process was used for Figure 3 in order to remove the possibility of this affecting the results.

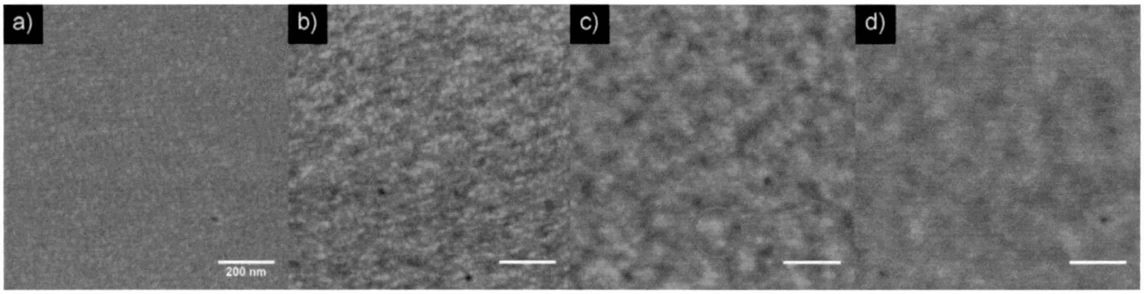

Figure 3: Thermally-annealed P3HT:PCBM film (150°C for 10 minutes) imaged using all annuli of the CBS detector using a) 500 eV, b) 1 keV, c) 2 keV (with high-contrast region highlighted), and d) 3 keV primary electron beam landing energy

Our interpretation of Figure 3 is thus: at 500 eV, the beam penetrates only the top few nm of the film. It is likely therefore that the P3HT 'capping layer' will have a significant masking effect on the imaged morphology as a relatively high fraction of the signal will originate from this layer. Using a 1kV primary beam, the majority of the signal originates from below the capping layer, as such the blend morphology is observed more clearly. The imaged morphology also appears coarser, perhaps representing the nature of the morphology at greater depths. At 2kV, whilst contrast between blend phases is still clearly visible, the phase structure is far less defined and is coarser still than that in Figure 3b. This partly results from some of the BSE signal originating from depths greater than the size of one phase –the BSE signal at a point on the surface has been projected through both P3HT and PCBM material. Very few areas (example circled) still show strong contrast, representing pure phases with a large surface depth (ideal for OPV morphology[10]). Hence for the given film thickness and materials, a 2 keV beam appears well suited to probe the suitability of an OPV morphology in terms of pure phases extending over a large surface depth. Using a 3 keV beam landing energy, the contrast has almost disappeared apart from some larger-scale features that we hypothesize could represent film thickness variations that alter the proportion of the BSE signal originating from the silicon substrate. Further work is required to develop these conclusions, and verify that this data is indeed reflective of the blend morphology as a function of depth. However we believe our study shows that film morphology variation can be explored by varying the landing energy of the primary beam.

4.1. Imaging considerations, optimisation of images at larger primary beam landing energies

We have found that beam optimization (i.e. stigmators, focus) can be challenging when using this technique, but if proper care is taken the image quality obtainable, in terms of signal-to-noise ratio and resolution, is more than adequate for our purposes at low-kV. We do note however that some image distortion can be found in the data at higher magnifications (field widths <1um).

At beam energies above 1keV, the image signal spreads from the A to the B annulus on the CBS detector, an effect confirmed by modeling [11] of the BSE emissions from P3HT, shown in Figure 4a. Here we plot the relative BSE signal across the angular range for the CBS detector. It is observed that the BSE signal using a 3keV landing energy is more evenly spread across the CBS detector's acceptance range than at 1keV. The effect is most clearly observed in the image data in Figure 4b - the best results in this energy range are achieved by combining the signal from the A and B annuli.

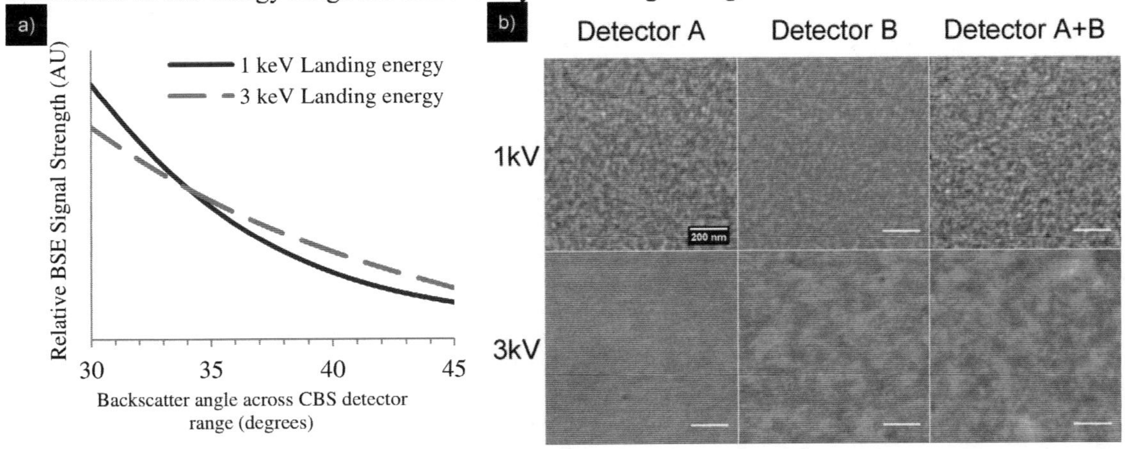

Figure 4: a) Spread of BSE signal across CBS detector at 1 keV and 3 keV; b) Variation in signal intensity between CBS detector annuli at 1 keV and 3 keV, subsequent optimization combining detectors A and B. Sample is a thermally annealed P3HT:PCBM blend, not plasma cleaned.

5. Conclusions

In conclusion, we have a demonstrated a low-kV BSE imaging technique that we believe enables a rapid overview of OPV blend morphologies. By employing stage biasing to produce images with the required clarity and resolution using a solid-state BSE detector, we have been able to map nanoscale phase variation in a P3HT:PCBM blend film. The technique appears capable of displaying the key morphological differences between P3HT:PCBM blends both as-cast and subject to a thermal anneal. Most importantly, we believe that there is genuine potential for probing some aspects of the sub-surface morphology without the need for sectioning techniques.

References

[1] Darling S B and You F 2013 *RSC Adv.* **3** 17633–48
[2] Liu Y, Zhao J, Li Z, et al 2014 *Nat. Commun.* **5** 5293
[3] Nelson J 2011 *Mater. Today* **14** 462–70
[4] Masters R C, Pearson A J, Glen T S, et al 2015 *Nat. Commun.* **6** 6928
[5] Phifer D, Tuma L, Vystavel T, et al 2009 *Micros. Today* **17** 40–9
[6] Drouin D, Couture A R, Joly D, et al 2007 *Scanning* **29** 92–101
[7] Joy D C 2002 *J. Microsc.* **208** 24–34
[8] Pfannmöller M, Flügge H, Benner G, et al 2011 *Nano Lett.* **11** 3099–107
[9] Huang Y, Kramer E J, Heeger A J, et al 2014 *Chem. Rev.* **114** 7006–43
[10] Moon J S, Lee J K, Cho S, et al 2009 *Nano Lett.* **9** 230–4
[11] Dapor M 2003, *Electron-Beam Interactions with Solids*; *Springer Tract Mod. Phys.* (Springer).

Separating topographical and chemical analysis of nano-structure of polymer composite in low voltage SEM

Q Wan[1], R A Plenderleith[1], M Dapor[3], S Rimmer[2], F Claeyssens[1], C Rodenburg[1]

[1]Department of Material Science and Engineering, University of Sheffield, Western Bank, Sheffield, S10 2TN, UK
[1]Department of Chemistry, University of Sheffield, Western Bank, Sheffield, S10 2TN, UK
[3]European Centre for Theoretical Studies in Nuclear Physics and Related Areas (ECT*-FBK) and Trento Institute for Fundamental Physics and Applications (TIFPA-INFN), via Sommarive 18, I-38123 Trento, Italy

Email: qwan2@sheffield.ac.uk

Abstract. The possibility of separating the topographical and chemical information in a polymer nano-composite using low-voltage SEM imaging is demonstrated, when images are acquired with a Concentric Backscattered (CBS) detector. This separation of chemical and topographical information is based on the different angular distribution of electron scattering which were calculated using a Monte Carlo simulation. The simulation based on angular restricted detection was applied to a semi-branched PNIPAM/PEGDA interpenetration network for which a linear relationship of topography SEM contrast and feature height data was observed.

1. Introduction

Due to the strong dependence of functional properties of polymer nano-composites on nano-morphological features, the characterization of nano-scale morphology is vital in the field of polymer nano-composite materials. Among all the techniques developed to provide morphological information, the SEM techniques processes several advantages: the simple sample preparation, contact free measurement, reasonable resolution in surface imaging and convenient scanning of a relatively large area comparing to other microscopies such as Transmission Electron Microscopy (TEM)[1] and Scanning Probe Microscopy (SPM)[2].

Low-voltage (LV) SEM was developed in order to overcome the radiation damage on beam-sensitive samples, still avoiding the need of conductive coatings[3]. Earlier application of LVSEM to materials such as regenerated silk fiboin nanofibre[4] and bone grafting[5] has proved to be an efficient technique. However in the condition of low-voltage SEM the energy of the backscattered electrons (BSE) may drop to similar levels of the secondary electrons (SE), and the normal separate detection of the SE and BSE may not be applicable. Thus effort to achieve separated topography and chemical information from SEM image contrast were made by the introduction of the theories of electron scattering in the imaging process. As shown in the studies of natural fibre[6] and metal[7] samples this electron kinetic based detection provided a potential extraction of topography and chemical contrast. The recent study on the photovoltaic materials has pushed this detection technique further to predictable material contrast mapping and identification[8].

The distribution of the scattered electrons may be analyzed for SE or BSE in several aspects such as the kinetic energy, the impact depth and the scattering angle. Using a concentric backscattered detector which can potentially detected ~50 eV signal[9], we have chosen the scattering angle of the backscattered electrons as the main factor in our separation of topographical and chemical information.

2. Experimental

The 0%, 5%, 10%, 20% and 30% extracted PNIPAM branched PNIPAM/PEGDA interpenetration network sample (the synthesis process is described in reference[10]) was treated with a Diener Zepto version B plasma cleaner. The gas input was air, and etching condition was set to 840 s with a power of 100 W. The plasma chamber was ventilated once during the etching to avoid overheating the sample surface. The AFM height data were acquired with a Dimension 3100 AFM in tapping mode. The PNIPAM/PEG samples were imaged by a FEI NOVA 450 SEM in low-voltage conditions (<5 kV accelerating voltage) and the spot size was set to 2 in all images. SEM images were taken using a concentric backscattered detector with 4000 V beam deceleration field in working distance ~4 mm. The applied detector segment and the accelerating voltage are listed for individual images.

3.Result and Discussion

According to the definition of contrast, the simulated image BSE contrast, C_s, can be described by the equation 1, as follows:

$$C_s = \frac{I_s - I_s'}{I_s'} \tag{1}$$

where I_s is the simulated BSE signal of the sample PNIPAM, and I_s' is the simulated BSE signal of the background (PEGDA matrix).
The calculation of the simulated BSE signal, I_s, is based on the equation

$$I_s \propto \eta \sum_{\theta_1}^{\theta_2} P_{elastic}(\theta) \tag{2}$$

where $P_{elastic}$ is probability of electron elastic scattering at angle θ, θ_1 and θ_2 are respectively the minimum and the maximum of the range of scattering angles for a given detector segment, and η is the backscattering coefficient of the respective material.

η is calculated by a Monte Carlo code and $P_{elastic}$ is computed by using the Mott theory (details of the Monte Carlo method and of the Mott theory are described in reference[11]). The CBS detector was installed at the end of the objective lens pole piece. The angular range of each CBS detector segment was calculated by the measurement of inner segment size and by literature data[9]. The diameter of the inner segment A is 4 mm and each segment is 0.8 mm wide. The working distance as set on the microscope was kept close to 4mm.

For our imaging condition of 700 V primary beam we found to be η=0.135 for PNIPAM and η=0.118 for PEGDA. The angular probability on each detector segment is shown in figure 1(a).

The comparison of simulation (C_s) with the experiment (C_e) in figure 1(b) shows that the backscattered electron signal dominates in the inner segments, especially in segments B and C, which allows the chemical mapping with BSE using these segments.

The simulation also revealed that the BSE signal on the outer segment D on the CBS detector is very weak for materials using small working distances. C_s in segment D at 4 mm working distance would be ~ -1%. This means that PNIPAM is darker than PEGDA matrix. However the C_e in segment D is ~ 21%. Due to the very weak BSE contribution to the signal in segment D in the simulation, we expect that C_e in the segment D, stem mostly from SE and thus provides topography information for our chosen experimental settings. However we note that the electron distribution is sensitive to the working condition of the microscope such as working distance and beam deceleration.

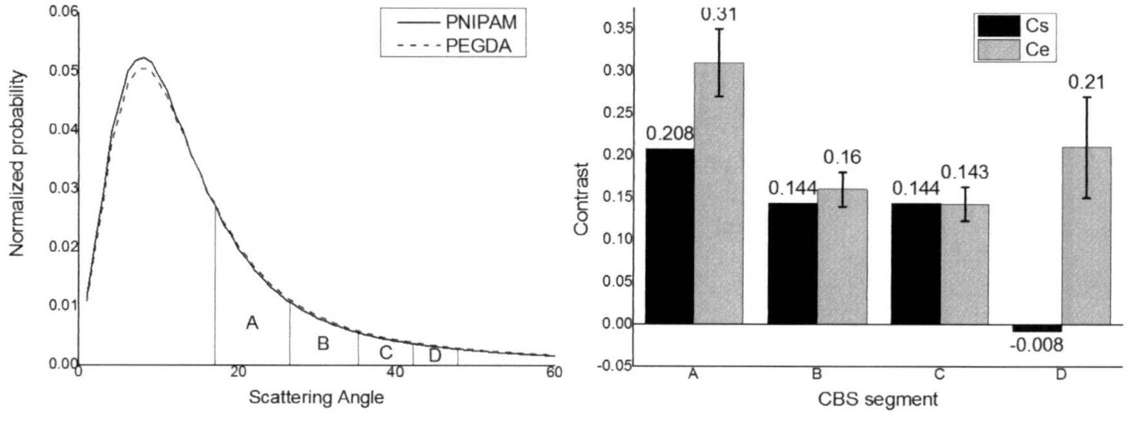

(a) (b)

Figure 1: The angular probability of backscattering angle (a) and the C_s and C_e in segments A-D (b) for a 20% etched PNIPAM/PEGDA sample imaged with 700 V primary beam energy and 4mm working distance.

We have imaged the plasma etched and un-etched PNIPAM/PEGDA samples with segment D at primary energy 700 V and working distance 4 mm, and compared the contrast of topographic features including the PNIPAM particle and PEGDA matrix phase separation (Figure 2(a)) to AFM height data (Figure 3). We note that C_e of objects on the segment D images has a rather linear relationship with the respective height data measured by AFM. The PNIPAM particle data is almost proportional while the PEGDA phase separation deviates only slightly below 30 nm height (Figure 2(b)). We find that a linear relationship is valid in height range of 30-100 nm for our objects with <300 nm diameter.

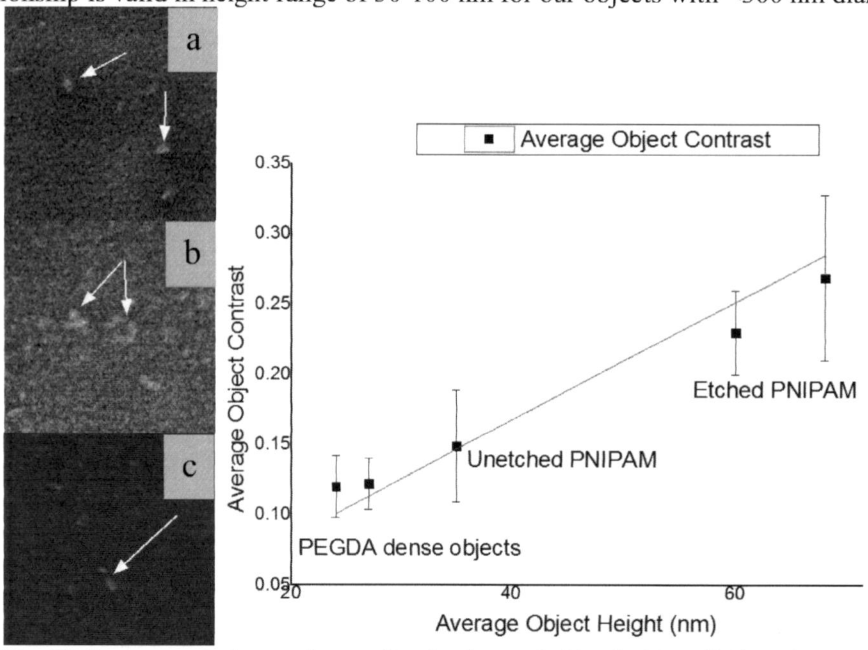

Figure 2: The CBS segment D image (normalized enhanced C_e) of 100 W/840 s plasma etched (a and b) and un-etched (c) PNIPAM particles and the average contrast vs average height of different objects..The field of view for SEM image is 2x2 µm. The PNIPAM particles are pointed out with arrows.

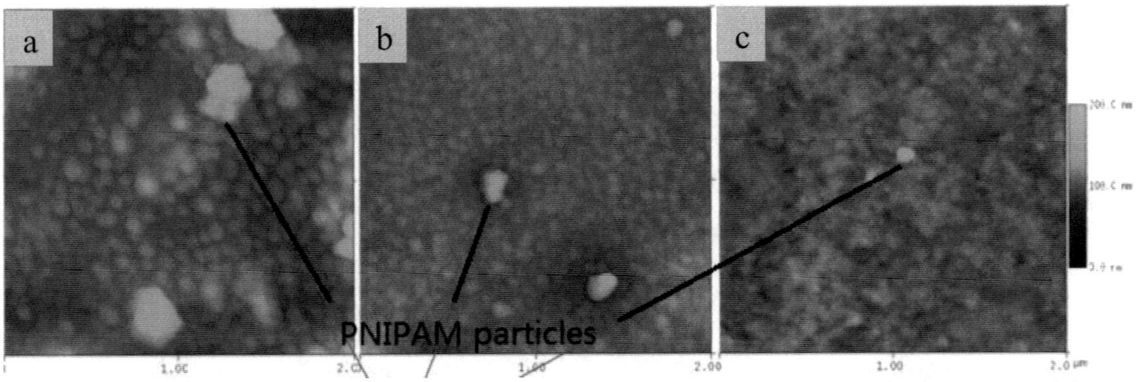

Figure 3: The AFM image of of 100 W/840 s plasma etched (a,b) and un-etched PNIPAM particles(c). The field of view in each image is 2x2 μm, the height range is 0-200 nm.

The deviation in the PEGDA phase separation feature may be due to the larger error in the measurement of smaller features by both AFM and SEM. More data on other materials should be collected to test the wider applicability for the linear height-contrast relationship. We note the peak in elastic scattering angular probability curve will shift to the smaller angles as the material becomes denser. Materials denser than our sample tend to have lower BSE signal in outer segment D. This indicates that our height mapping may still be reasonable for materials denser than our polymers.

4. Conclusion

The distribution of backscattered electrons was simulated and the separation of chemical and topographical information in LVSEM was achieved through the controlling of angular range of detection in our experiment. We found an approximately linear relationship between contrast C_e in CBS D segment and the AFM height data for two polymers. Such a relationship therefore could be used for contact free direct height mapping of nano-topographical features in LV-SEM once a calibration curve is recorded. Simulations suggest the wider applicability to denser materials.

References

[1] Danino D, 2012 *Curr Opin Colloid Interface Sci*, **17** 316
[2] Vansteenkiste S O, Davies M C, Roberts C J, Tendler S J B, Williams P M, 1998 *Prog Surf Sci*, **57** 95
[3] Vezie D L, Thomas E L, Adams W W, 1995 *Polymer*, **36** 1761
[4] Mhuka V, Dube S, Nindi M M, Torto N, 2013 *Macromol Res*, **21** 995
[5] Benezra Rosen V, Hobbs L W, Spector M, 2002 *Biomaterials*, **23** 921
[6] Rasch R, Stricher A, Truss R W, 2013 *J. Appl Polym Sci*, **131**
[7] Nagoshi M, Aoyama T, Sato K, 2013 *Ultramicroscopy*, **124** 20
[8] Masters R C et al, 2015 *Nat Commun*, **6** 6928
[9] Sakic A et al, 2011 *Micro Technology and Micro Devices SAFE, 1-4*
[10] Plenderleith R, 2014 *thesis* University of Sheffield
[11] Dapor M, 2014 *Transport of Energetic Electrons in Solids*, Springer. Tr. Mod. Phys., Vol. 257 (Springer, Berlin)

Exploring backscattered imaging in low voltage FE-SEM

P Lewis[1], S Micklethwaite[1], J Harrington[1], M Dixon[2], R Brydson[1] and N Hondow[1]

[1] School of Chemical & Process Engineering, University of Leeds, Leeds LS2 9JT, UK
[2] Hitachi High-Technologies Europe, Maidenhead SL6 8YA, UK

pmpml@leeds.ac.uk (P Lewis) n.hondow@leeds.ac.uk (N Hondow)

Abstract. Contrast levels in backscattered SEM images were investigated, utilising stage deceleration for low voltage imaging and also electron energy filtering. Image contrast variations are explained via use of Monte Carlo simulations which can predict the optimum accelerating and filter voltages for imaging complex sample mixtures.

1. Introduction

Low voltage scanning electron microscopy (SEM) is advantageous due to both the increased surface sensitivity for imaging and the reduction of specimen charging [1]. The accelerating voltage of a SEM is normally limited to a value such as 0.5 or 1 kV whilst preserving acceptable image resolution, however lower voltages can be achieved by using approaches such as deceleration, where a negative potential applied to the sample stage allows electrons to land at much lower voltages.

In addition to secondary electron (SE) imaging, recording the backscattered electron (BSE) signal can provide useful information which is normally related to the average atomic number at the sample position [2]. However Müllerová and Frank have shown that at acceleration voltages below 2 kV the dependence between backscattered intensity and atomic number (Z) breaks down [3] and light elements more strongly backscatter electrons than heavier elements often leading to a contrast inversion in images [3]. It is also possible to enhance contrast at low acceleration voltages by selectively filtering the energy of emitted electrons thereby permitting the detection of specific energy back scattered electrons. Whilst not as common as energy filtering in a transmission electron microscope, energy filtered SEM has shown application in the imaging and analysis of semiconductors [4,5], in particular in reducing the influence of surface contamination layers in SE images [5]. In further experiments imaging a thermally-oxidized carbon steel and a Zn-55%Al coating, specimen conditions were kept constant whilst the bias voltage was varied [6]. It was found that image brightness decreased with increasing bias voltage and the brightness changed drastically around 0.5 kV for both specimens. Here we investigate the potential for low voltage backscattered imaging of complex mixtures of materials.

2. Experimental

SEM was conducted on a Hitachi SU8230 cold field emission (FE)-SEM equipped with an Oxford Instruments 80 mm^2 X-Max energy dispersive X-ray (EDX) detector. The Hitachi SU8230 FE-SEM is equipped with numerous detectors allowing the simultaneous recording of various electron signals and energies (figure 1). Imaging was conducted using incident beam energies of between 2 and 0.25 kV, with energies less than 2 kV achieved via deceleration, together with EDX analysis at 20 kV.

Three samples were examined; a mixed aluminium-gold (Al-Au) sample prepared by Au coating an Al stub; a mixed particulate sample of ~25 nm ceria (CeO_2) and ~25 nm titania (TiO_2) nanoparticles made by allowing a drop of each nanoparticle dispersion in methanol to dry on an Al stub; and finally a mixture of carbon nanotubes (CNTs) and TiO_2 nanoparticles deposited on an Al stub. The Al-Au and CeO_2-TiO_2 samples were imaged using a photodiode-backscattered electron (PD-BSE) detector at beam energies of between 1.5 and 0.25 kV, achieved via deceleration of an initial 2 kV electron beam (figure 1a). The CNT-TiO_2 sample was imaged at 2 kV using the upper (figure 1 b) and top (with and without filter; figure 1c) detectors.

The Monte Carlo simulation software Casino Flight Simulator (Version 2.4.8.1) was used to calculate values of backscatter co-efficient for electrons hitting the surfaces of samples of different composition. Each simulation used the Mott cross section for interpolation with one million electrons. Simulations were run on Al and Au at acceleration voltages of 0.1 to 1.0 kV, in 0.1 kV steps, and on CeO_2 and TiO_2 at acceleration voltages of 0.1 to 1.5 kV, in 0.1 kV steps.

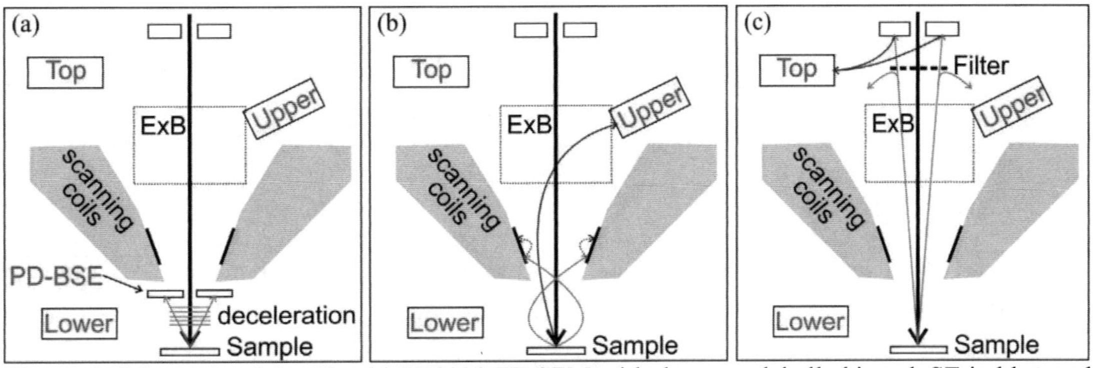

Figure 1. Schematic of the Hitachi SU8230 FE-SEM with detectors labelled in red, SE in blue and BSE in green. (a) PD-BSE detector to form BSE images. (b) Upper detector used to form SE images. (c) Top detector used to produce BSE images, the energy of which can be filtered.

3. Results and Discussion

3.1. Backscattered electron imaging at low voltages

The Hitachi SU-8230 FE-SEM used in this study can operate at a minimum accelerating voltage of 0.5 kV. To operate at lower voltages, stage deceleration is used. While this can allow imaging as low as 10 V, we have used an initial operating voltage of 2 kV which is then dropped to an effective voltage of between 1.5 and 0.25 kV by applying a potential of between 0.5 and 1.75 kV to the sample stage. The advantages of deceleration are that the electron beam retains the lower aberration and higher brightness associated with higher acceleration voltages but also has the smaller interaction volume of lower voltages, this along with the focusing effect caused by the magnetic field which is created around the sample by the deceleration voltage greatly increases the resolution of the system.

The backscattered signal can be detected using the PD-BSD and allows the formation of images that permit the discrimination between materials of different atomic weights. When examining the Al-Au sample at 1 kV (figure 2a), there is a darker level of contrast in the lower atomic number Al (Z=13) relative to Au (Z=79), as would be expected by the standard Z-contrast. This contrast is consistent at 0.5 kV (figure 2b), however at 0.25 kV (figure 2c) there is a contrast inversion, with the Au now darker than the Al. Confirmation of element location is provided by the EDX map (figure 2d), and in addition images exhibit more topographical contrast as the deceleration voltage is increased.

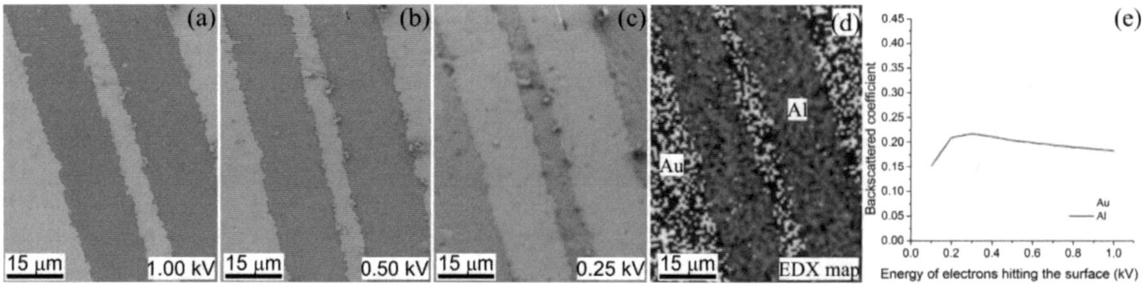

Figure 2. SEM images and analysis of the Al-Au sample. As the voltage is reduced a contrast inversion is observed between Al and Au. Backscattered electron images recorded at (a) 1.00 kV, (b) 0.50 kV and (c) 0.25 kV. (d) EDX map showing the distribution of Al and Au. (e) Plot of the predicted backscatter coefficient of Al and Au against voltage; data generated using Monte Carlo modelling. The intersection point, and therefore theoretical contrast inversion point, is at 0.41 kV.

To further understand this contrast inversion, Monte Carlo simulations were conducted to determine the backscatter coefficients for each element at the appropriate effective accelerating voltages. Using appropriate conditions for the simulations, a contrast inversion is predicted to occur at 0.41 kV for Al and Au (figure 2e), which matches the experimental results (figure 2a-c).

The ceria-titania nanoparticle sample was examined under similar conditions. This proved to be challenging, as the size and shape of the nanoparticles was remarkably similar. Images recorded at 1.5 kV (figure 3a) show distinctly different areas, with darker contrast TiO_2 nanoparticles (for Ti, Z=22) and brighter CeO_2 nanoparticles (for Ce, Z=58). At lower voltages, such as 0.25 kV (figure 3b) the contrast was very similar between the different nanoparticles. Monte Carlo simulations (figure 3c) predict that a contrast inversion would take place at 0.26 kV; however the contrast levels between the two materials remain very similar, not allowing for identification of each nanoparticle type via image contrast alone at very low voltages.

Figure 3. SEM images and analysis of a CeO_2-TiO_2 mixture. As the voltage is reduced the contrast difference observed between CeO_2 and TiO_2 changes. Backscattered electron images recorded at (a) 1.00 kV and (b) 0.25 kV. (c) Plot of the predicted backscatter coefficient of CeO_2 and TiO_2 against voltage; data generated using Monte Carlo modelling.

3.2. Energy filtering and low-loss backscatter electrons

The CNT/TiO_2 sample on an Al stub was examined at 2 kV. Figure 4a is a SE image formed using the upper detector where the low angle BSEs are attracted to the positively charged bias plates (figure 1b) and, while this does cause low energy SE emission, this is pulled back to the plates, and consequently the image is a pure SE signal giving highest resolution and surface sensitivity.

The backscatter top detector with the high-pass filter (figure 1c) permits filtering of the signal; it is possible to image using the entire backscattered signal or only the elastically scattered backscattered electrons, the so called low-loss backscatter signal.

Figure 4b does not use the filter, and here the low Z CNTs show similar contrast to the higher Z Al substrate beneath. Here the elastically and inelastically scattered BSE signal is collected, the escape depth of which is greater than the thickness of the CNT and close to the total interaction volume, such that the contrast consists of a convolution of signals from the Al substrate and the CNT together. The contrast cannot be interpreted in the conventional way expected of a backscatter signal.

Figure 4c does use the top filter such that only low-loss BSE signal is acquired. The signal sampling depth is constrained to approximately the mean free path of the backscatter electrons (since only elastically scattered electrons are being collected and backscatter electrons with energy losses emerging from deeper within the sample are rejected). Thus, the signal becomes sufficiently surface sensitive to represent the correct backscatter contrast of the materials involved and the image can be more directly interpreted. This technique may be used to provide high surface sensitivity compositional imaging without need to reduce the interaction volume or enter the regime of potential contrast inversions.

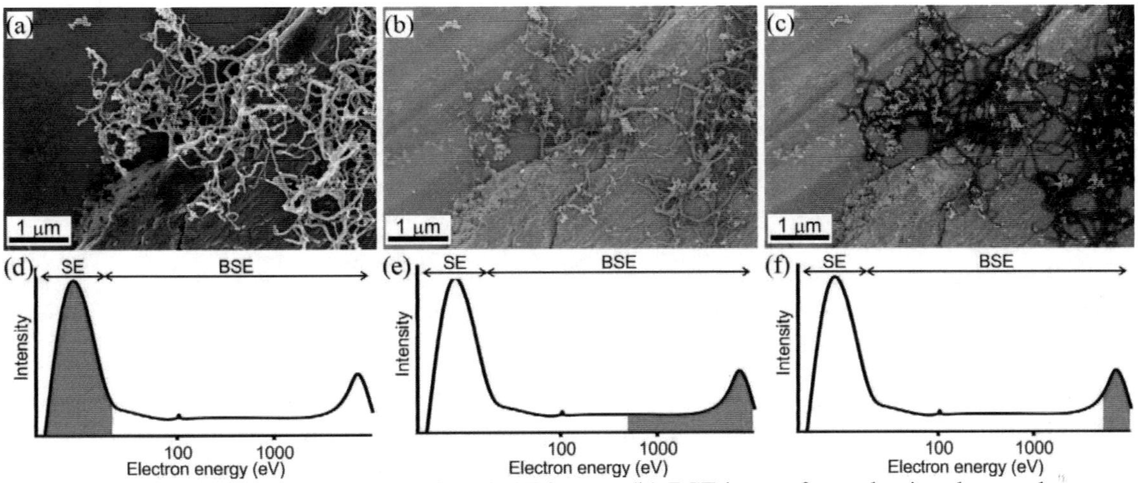

Figure 4. Images of CNT/TiO_2 sample. (a) SE image. (b) BSE image formed using the top detector without the filter. (c) BSE image formed using the top detector with the filter. (d)-(f) Schematic of electron energy spectra associated with images (a)-(c) respectively.

4. Conclusions

These results indicate how low voltage SEM is an effective tool for imaging materials with different average atomic numbers. The use of BSE imaging at low voltages, with and without energy filtering, can provide interesting contrast levels in images, some of which do not follow the expected Z-contrast. The use of energy filtering enables the surface sensitivity of backscatter imaging to be enhanced and can lead to more directly interpretable images. Monte Carlo simulations are a very effective tool for predicting when contrast inversion will occur, and can both explain the contrast levels observed and could be used to determine the most appropriate accelerating voltage to use when examining a mixture of materials.

References

[1] Müllerová I 2001 *Scanning* **23** 379
[2] Goodhew P J, Humphreys J and Beanland R 2001 *Electron Microscopy and Analysis, Third Edition* (London: Taylor & Francis)
[3] Müllerová I and Frank L 2004 *Scanning* **26** 18
[4] Rodenburg C, Jepson M A E, Inkson B J, Bosch E G T and Humphreys C J 2010 *J. Phys. Conf. Proc.* **241** 012074
[5] Tsurumi D, Hamada K and Kawasaki Y 2012 *Jpn J Appl Phys* **51** 106503
[6] Nagoshi M, Aoyama T and Sato K 2013 *Ultramicroscopy* **124** 20

Cascade Model of Noise Formation in Electron Detectors

S Shofman[1], E Cheifetz[1], A Weingarten[1], M Berse[2]

El-Mul Technologies Ltd., 12A Hamada St., Rehovot 7670315, Israel
Raith GmbH, Konrad-Adenauer-Allee 8, 44263 Dortmund, Germany

E-mail: semyon.shofman@el-mul.com

Abstract: Signal and noise evolution in electron detectors are analysed using a cascade model. For each cascade, the required parameters can be obtained experimentally or by simulation. The model allows predicting the noise properties of a detector, in order to identify and improve the most important stages in the signal formation chain, or in return to evaluate the performance of unknown stage. Experimental results from Raith ionLINE tool are presented, demonstrating good agreement with the model.

1. Background

Signal to Noise Ratio (SNR) is one of the most critical properties of any imaging system. In electron microscopy, it is governed by probe beam shot noise and by statistical nature of signal formation within the electron detector. The image formation is a chain of stochastic transformations of the primary beam: Secondary Electron (SE) or Backscattered Electron (BSE) emission from the sample, detection of part of the emitted electrons, and conversion of the detected electrons to electronics signal. The later can be either direct multiplication of the collected electrons, as in solid state detectors, or transforming the electrons to light using scintillator, use a light guiding system and then convert it back to electronic signal which is multiplied. All detectors use additional stage of electronics which is not taken into account in this analysis.

Due to the statistical nature of the transformations in the chain, each step adds noise to the previous stage. In this paper we present an analysis for the most common scintillator-based chamber detector, but the model can be used for analyzing practically any detection system. The treatment presented here is an improvement over previous ones[1][2], by providing a detailed and universal calculating method, taking into account more factors in the noise formation

2. Noise factor calculation

The detector Noise Factor NF is the measure of SNR deterioration along the signal formation chain, and is defined as the ratio between input SNR to output SNR. Each transformation is considered as cascades with statistical gain, which has mean value m and standard deviation s. Denoting $\gamma = m / s$ the formula for NF calculation can be expressed as:

$$NF = \sqrt{1 + \frac{1}{\gamma_1^2} + \frac{1}{\mathrm{m}_1 \gamma_2^2} + \frac{1}{\mathrm{m}_1 \mathrm{m}_2 \gamma_3^2} + \ldots}$$

This formula is derived from a common expression for the noise of multi-cascade systems [3] assuming shot noise (Poisson process) in the input of the first cascade. In the electron detection system, this would be the probe beam shot noise.

Signal formation chain includes Poisson processes for which $\gamma^2 = m$, binary processes for which $\gamma^2 = \frac{m}{1-m}$, and more complex processes for which γ is more complicated and in some cases, should be obtained experimentally.

Applying the model to signal formation in Everhart-Thornley (ET) chamber detector in an ion beam tool, we consider the following cascades:

- Secondary electron (SE) emission from sample
- Collection of electrons emitted from sample to scintillator
- Photon generation within scintillator
- Light transmission from scintillator to Photo-Multiplier Tube (PMT)
- Photoelectron generation on PMT photocathode
- Electron multiplication in PMT

SE emission from sample and electron multiplication on each PMT dynode are Poisson processes, for which $m_i = \sigma_i$, where σ_i is SE yield for each stage. Electron collection, light transmission and photoelectron generation are binary processes since in these cases, the electron, photon or photo-electron either reaches the next stage in the chain or not. These stages satisfy $m_i = \eta_i$, where η_i is electron detection efficiency, light collection efficiency and photocathode QE respectively. Photon generation within scintillator is a more complicated process since it involves the location of light formation, its distribution, light propagation in scintillating material and in the substrate, and light extraction from the substrate. While it can be readily calculated for crystal scintillators, it becomes model dependent for granulated phosphorous materials, which are most commonly used. Therefore, for our analysis, we obtained the m and γ parameters experimentally.

3. NF calculation for ionLINE tool

The NF calculation was employed for the ETD detector developed by El-Mul for Raith ionLINE tools in order to evaluate and optimize the detector performance. The cascade stages with their m-values are shown in Table 1. SE yield on sample was taken from literature[4]. The electron collection efficiency was calculated by 3D electron trajectories simulation under the electric field which is calculated from the actual electrodes construction and their voltages. Light transmission efficiency was simulated by El-Mul, γ for the phosphor was obtained experimentally from its cathodoluminescence image taken in a SEM, and PMT manufacturer's QE and gain data were used.

Stage	Parameter	m-Value
1	SE from sample (for 35keV Ga ion beam on Al sample)	1.5
2	Electron collection efficiency	0.2
3	Photon generation in P47 scintillator	110
4	Light collection from scintillator to PMT	0.66
5	Photoelectron generation in PMT	0.13
6	SE from PMT dynodes	3.4

Table 1: m values for ETD process steps

The NF calculation results are shown in Figure 1. It can be seen that the SE generation and the electron collection efficiency to the detector are the most dominant NF contributors.

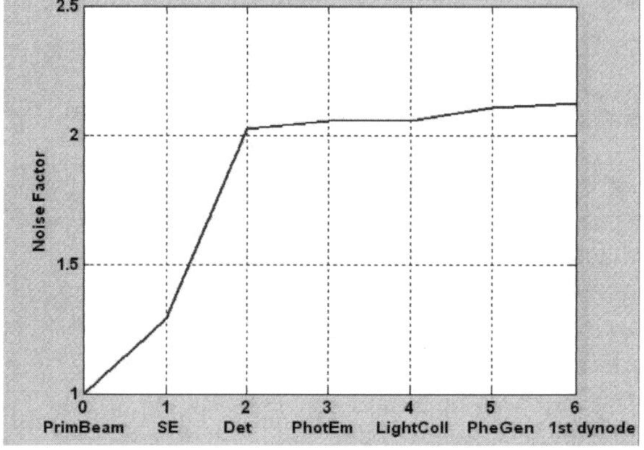

Figure 1: Noise factor evolution for ionLINE ETD

The images of a flat uniform Silicon sample were taken with both the original Raith and new El-Mul detectors, and their SNR were obtained by image processing. As Ga+ ion beam does not give backscattered particles from Si, only SE generated by ion beam are detected. For both detectors the NF calculated by the presented formula and obtained by image processing are in good agreement.

Detector	NF Calculation	Experimental NF
Original Raith detector	4.81	4.76
El-Mul detector	2.14	2.25

Table 2: comparison of calculated and measured NF values

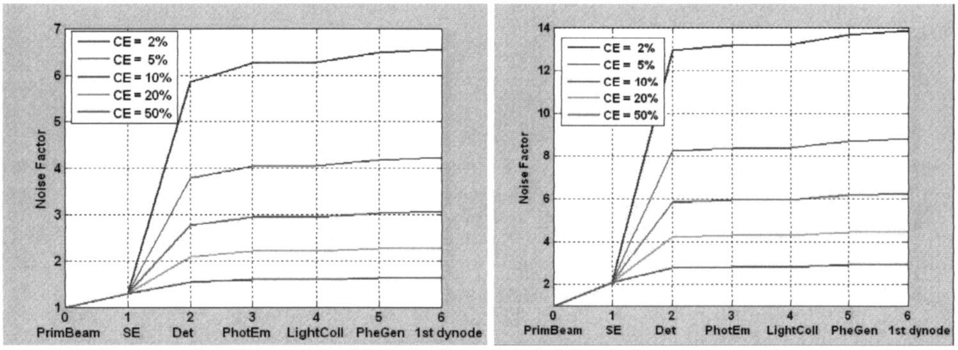

Figure 2: NF values for SE yield of 1.5 (left) and 0.3 (right) for various electron collection efficiencies

4. Electron and light collection efficiency effect

Since the electron collection efficiency is the most contributing part for the detector simulated here, the NF was calculated for various values in order to evaluate the possibility for improvement. The calculation results, shown in Figure 2 (left side), demonstrate that increasing the collection efficiency from 20% to 50% will result in SNR improvement by ~20%. Generally, the electron collection efficiency is limited by the available space around the pole piece, limited by various elements in the chamber, and such improvement in collection efficiency is not possible. However, in some cases, it is possible to add another detector that may almost double the efficiency. Figure 2 also shows that also in the case of low SE yield from the sample, the CE plays a major role in determining the SNR.

Although in the ETD case, the influence of the cascade stages following the electron detection are negligible, in some cases they can become major factor. For example, for In-Lens detectors or when space is very limited and the light guide becomes complicated, the light transmission efficiency might become a dominating factor. Figure 3 demonstrates that when light collection efficiency becomes of the order of 15% or lower, it starts to play an important role in the noise factor. Only a small fraction of the photons reach the PMT and the number of photo-electrons at the PMT photocathode becomes very small with poor statistics. The contribution of the photo-electron generation then becomes a major factor. This in turn, can be improved by using a more efficient PMT.

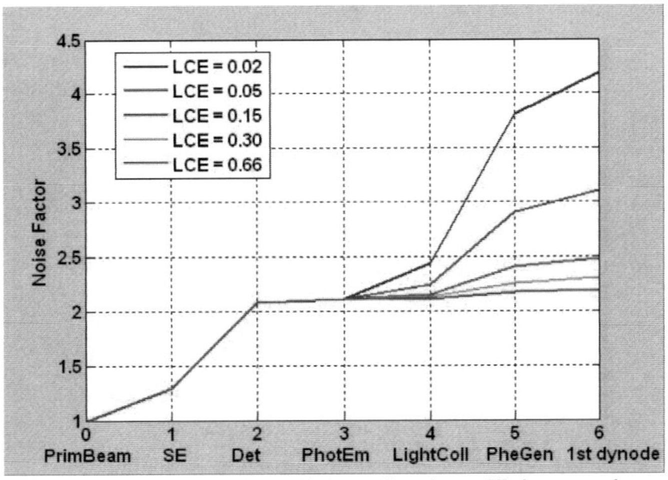

Figure 3: NF for various light collection efficiency values

5. Summary

A general model for deriving output SNR of electron detectors in electron or ion beam imaging tools was presented. The SNR calculated by the model was compared with experimental SNR for the case of ETD in a Ga+ beam tool and the results were in good agreement which also confirms the accuracy of the electron collection efficiency simulations. The model can be extended to many types of electron/ion detecting systems e.g. detectors with ion/electron convertor, with MCP or SSD based. Applying the model during detector design enables to identify the critical stages where noise formation should be reduced.

References

[1] Baumann W and Reimer L 1981 *Scanning* **4** 141-151
[2] Timischl F, Date, M and Nemoto S 2011 *Scanning.* **33** 1-8
[3] Vartsky D, Feldman G, Mor I, Goldberg MB, Bar D and Dangendorf V 2009 *J. Instrum.* **4** P02001
[4] Alonso EV, Baragiola RA, Ferron, J, Jakas MM and Olivaflorio A 1980 *Phys. Rev. B* **22** 80-87

Atomic-scale insights into 1D and 2D nano-materials

U Bangert[1], W Pierce[2], C B Boothroyd[3], M Migliorato[4], C-T Pan[2], A J Harvey[1], D M Kepatsoglou[5], Q M Ramasse[5]

[1]Department of Physics and Energy, University of Limerick, Limerick, Ireland,
[2]School of Materials, The University of Manchester, Manchester M13 9PL, U.K.
[3]Ernst Ruska-Centre for Microscopy and Spectroscopy with Electrons and Peter Gruenberg Institute Juelich Research Centre, D-52425 Juelich, Germany
[4]School of Electrical and Electronic Engineering, The University of Manchester, Manchester M13 9PL, U.K.
[5]SuperSTEM Laboratory, STFC Daresbury Campus, Daresbury WA4 4AD, United Kingdom
e-mail: Ursel.Bangert@ul.ie

Abstract. Atomic resolution imaging and narrow-energy spread spectroscopy in aberration corrected (scanning) transmission electron microscopes, in combination with DFT modelling has made it possible to uncover atomic-scale morphology, defect constellations, lattice impurities and ad-atoms in nano-materials, as well as revealing their influence on the surrounding bandstructure. Using atomic-scale imaging, EEL spectroscopy and EFTEM, we address issues beyond the more common investigations of their atomic lattice structure. We focus on the demonstration of (i) ripples in graphene and on effects of (ii) metal ad-atoms as well as of (iii) controllably introduced impurities -via low energy ion implantation- in both, graphene and carbon nanotubes, on the electronic band structure. We demonstrate the creation of a new feature with collective charge carrier behaviour (plasmon) in the UV/vis range in graphene and carbon nanotubes via EEL spectrum imaging and EFTEM, and support this with dielectric theory modelling.

1. Introduction. Graphene and carbon nanotubes have been intensively investigated via electron microscopy methods [1-3] regarding their atomic structure, and peculiarities of the latter introduced by changes of morphology, e.g., at the edges of carbon sheets or through impurities [4,5]. Less frequently addressed phenomena concern the local topography (flatness), especially of suspended graphene, and its implications on electronic structure. The effect of impurity incorporation, albeit through physio-chemical adsorption, bonding or integration into the carbon lattice on bandstructure in the eV regime, i.e., in the UV/vis energy excitation regime is a further, less actively researched area; the research focus is predominantly on terahertz and infrared physics/ applications with regards to opto-electronics and plasmonics. Controlled impurity integration, leading to bandstructure tailoring other than by electric gauging or chemical functionalization, is a field of interest, as chemical methods have significant drawbacks due to lack of control, contamination and inferred secondary impurities, instability and site selectivity. Ion implantation, enabling flexible small-depth channel doping (like in semiconductor-technology) is a prospective method for non-chemistry reliant functionalisation and processing of 2D and 1D materials [6,7]. Here we present a selection of investigations concerning the above topics with the aim of demonstrating impact of nano-structure on electronic bandstructure phenomena in the UV/vis energy regime.

2. Experimental. *Sample preparation.* Graphene was produced via chemical vapour deposition (CVD) of CH_4 and H_2 precursor gasses on a copper foil substrate [8]. The graphene was then transferred onto Quantifoil TEM grids, using the 'wet transfer' method [9]. Freestanding graphene produced using this method was mainly monolayer, with occasional bilayer or multilayer regions. As with all graphene, residual hydrocarbon deposits were seen on the samples. Metals were deposited onto the graphene using electron-beam or thermal evaporation with an evaporation rate of 0.1 nm/s, resulting in metal deposits of 0.1 to 0.3 nm height.

Carbon nanotubes (CNTs) were doped directly with alkali and earth alkali metals (e.g. K, Na and Ba) and B via ion implantation at ~200 eV at the Surrey University Ion Beam Centre, and with B and Ag, under the same conditions, at the Salford University Low Energy Ion Implanter, all to doses of

10^{14}-10^{15} cm^{-2}. The implantation energy was chosen to ensure uniform doping over the volume of nanotubes containing multiple carbon sheets. Substitutional doping of graphene was achieved using ultra-low-energy ion implantation [6]. Energies of ~25 eV were chosen and implantation was carried out with the University of Göttingen mass selected ion beam deposition system [10, 11], directly into free standing graphene, at doses between 10^{14} and 10^{15} cm^{-2}. In support of the ultra-low energy ion-implantation experiments, simulations of the implantation process in graphene were previously performed in Göttingen using the Monte Carlo program SDTrimSP [10,12], suggesting doping levels of 1.5 to 5 at% [6, 10]. Previous HRTEM investigations indicate nitrogen doping levels ranging from 0.1 at% to 1 at% [6], and higher (few %) for alkali-ions in CNTs.

Data acquisition. Energy filtered TEM (EFTEM) imaging and energy loss spectroscopy (EELS) in conjunction with high resolution bright field microscopy (HREM) were carried out at the Ernst-Ruska Centre, Jülich, on an FEI Titan 50-300 Pico triple aberration corrected and monochromated TEM operated at 80 keV. The microscope is equipped with a Gatan UltraScan 4000 UHS CCD with 4k x 4k pixels. EFTEM images and EELS were acquired using a Gatan Quantum ERS image filter with fully 2nd and 3rd order and partially 4th order corrected prisms. EFTEM images were acquired with a slit width of between 0.4 and 1 eV and with energy step sizes of between 0.1 eV and 0.5 eV. EEL spectra were (i) extracted from EFTEM data cubes or (ii) directly acquired in scanning (STEM) mode. Monochromation of the electron beam enabled the FWHM of the zero-loss peak to be reduced to around 0.1 eV. Additionally, atomic resolution HAADF and Spectrum imaging were carried out on the Daresbury Nion UltraSTEM, with a spectrum resolution of 0.3 eV. Graphene HREM images were additionally obtained in the Warwick ARM 200F.

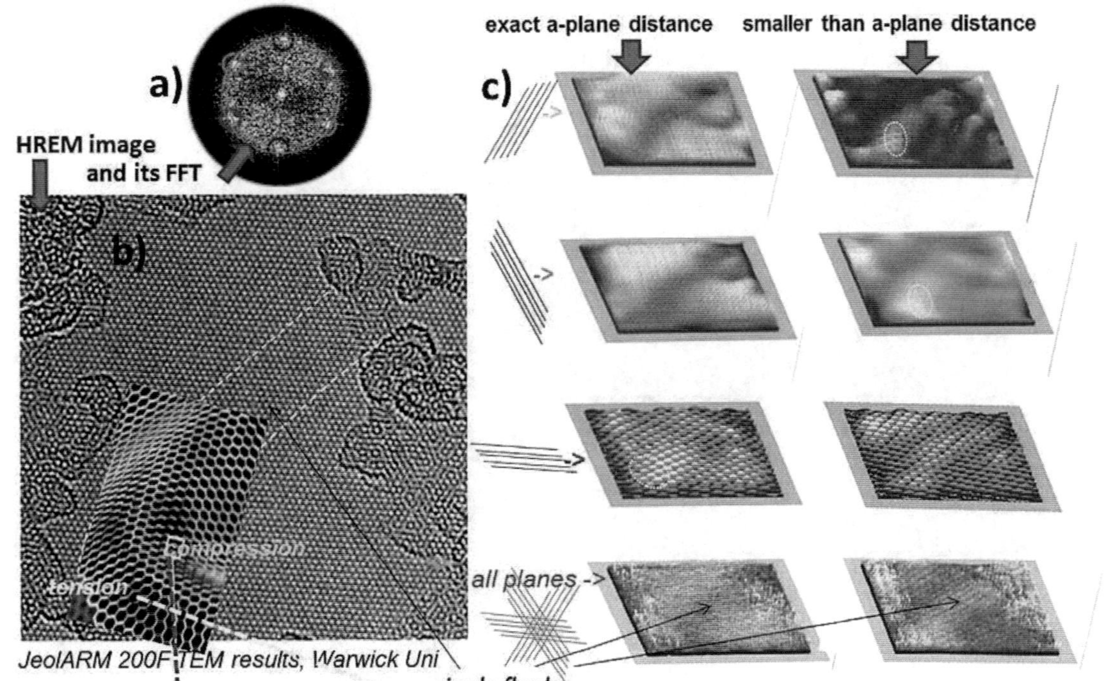

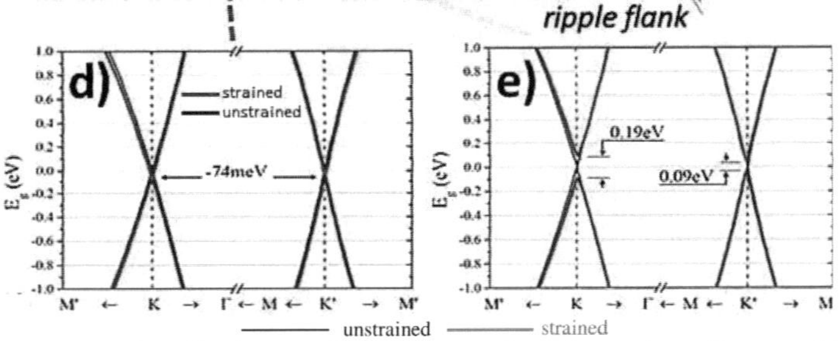

Fig. 1 a) Fourier transform of the HREM image of graphene in **b)**, overlaid on which is a cartoon of a ripple established in via the filtering method demonstrated in **c)**: the left-hand column contains inverse Fourier transform images after placing a filter at the exact reciprocal *a*-plane distance in the Fourier transform diffraction pattern in (a). This filters out unstrained, flat regions or inclined regions under tension. The right-hand column in (c) contains Fourier transform images after filtering smaller distances than the *a*-plane distance, highlighting inclined regions or regions under compression; **d)** bandstructure at the Dirac points, calculated by tight binding, predicting a bandgap of 0.19 eV at the K point of the Brillouin zone in regions of tensile strain; **e)** no bandgap opening occurring in regions of compressive or low (to zero) strain. However for compressive strain the Dirac point lies below the Fermi energy by 74meV.

Journal of Physics: Conference Series **644** (2015) 012021 doi:10.1088/1742-6596/644/1/012021

Data evaluation. Zero-loss peak (ZLP) extraction was carried out in Gatan Digital Micrograph (DM) by fitting a power-law background to the ZLP using a vacuum spectrum (obtained far from the sample) as a guide. An alternative approach, using the reflected-tail function in DM, was used for the EFTEM images recorded on the monochromated Titan Pico, with the monochromation achieving small and nearly symmetric tails of the ZLP. In some cases (on spectra obtained from metal- and B-doped CNTs) Richardson Lucy (RL) deconvolution was applied, bringing the spectral information down to 1 eV.

Calculations and modelling of low loss EEL spectra entailed calculations of the energy loss function and complex dielectric function of graphene and SWNT systems using the WIEN2K DFT code [13], aiding investigations of whether an excitation is of collective (plasmon) nature. The dielectric function was also extracted from experimental EEL spectra using the Kramers-Kroenig Analysis programme in Digital Micrograph (with a refractive index of around 2; a full description of the analysis method can be found in the Gatan Documentation, EELS Analysis, Copyright 2012 Gatan Inc.).

3. Results and Discussion.

(i) *Ripples.* Contamination-free areas in suspended graphene are not atomically flat, but have been shown to have static ripples [14,15], of the order of up to 10 nm in width and a couple of nm in height. Apart from direct observation via STEM [15] the morphology of ripples can be obtained via Fourier filtering of atomic resolution lattice images [14]. Figure 1a) shows an FFT of the area shown in fig.1b. A spot filter mask was then set to encompass regions around the diffraction spots at different reciprocal distances. Figure 1c (left-hand column) shows inverse FFT images after having set a spot mask at the exact *a*-plane distance (concentrically encompassing the diffraction spots) for the three *a*-planes in turn. The inverse FFT will then filter out exact *a*-plane distances/ bond lengths in flat regions, as well as bonds, which are longer (tension), but on inclined surfaces, resulting in a projected bond length that appears the same as that of the graphitic *a*-planes. Figure 1c (right-hand column)

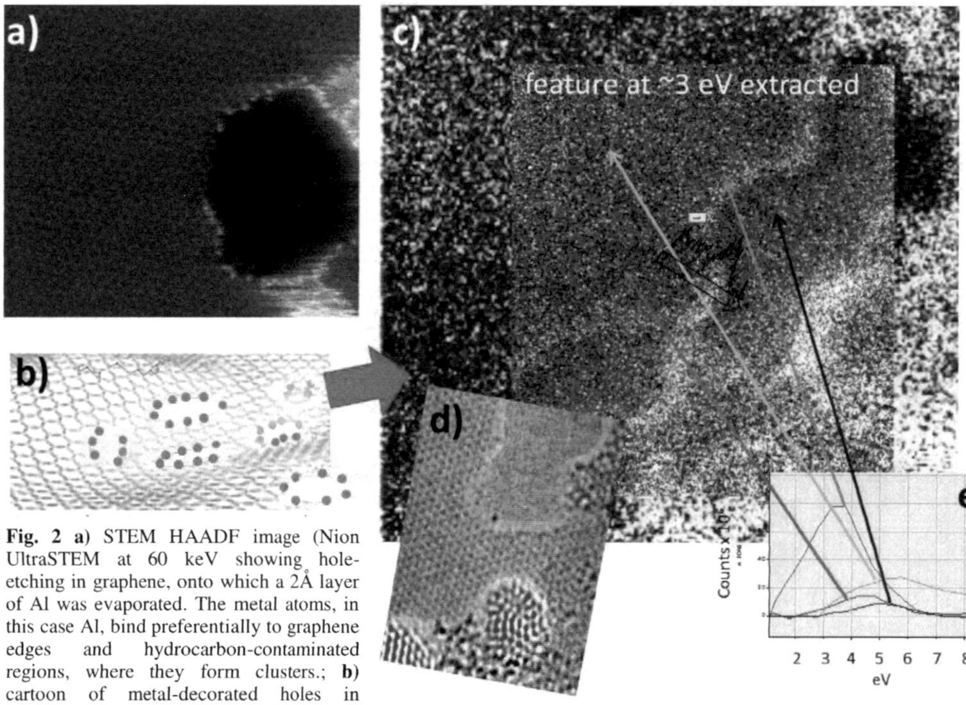

Fig. 2 a) STEM HAADF image (Nion UltraSTEM at 60 keV showing hole-etching in graphene, onto which a 2Å layer of Al was evaporated. The metal atoms, in this case Al, bind preferentially to graphene edges and hydrocarbon-contaminated regions, where they form clusters.; **b)** cartoon of metal-decorated holes in graphene, giving rise to enhanced plasmon emission, as experimentally observed; **c)** ZLP subtracted EFTEM image (blue and yellow colours) with overlaid EFTEM image after ZPL and π-plasmon subtraction (orange and blue colours) at the excitation energy of 3-4 eV, of Pd-dosed graphene, showing enhancement in the UV/vis regime at the graphene edge around a hole. An intensity profile along the black arrow across the hole's edge shows ~50% enhancement. The very bright intensity at the bottom of the EFTEM image stems from Pd crystals; **d)** overlaid structural (HREM) image, with dark dots overglossed by yellow sheen) representing Pd atoms; **e)** intensity of residual EEL signal after removal of the excitation in the π-plasmon region at 4.5 eV, revealing the new plasmon feature at slightly above 3 eV, only visible in the presences of Pd-atom decoration (frame width of EFTEM images ~15 nm).

shows inverse FFT images after setting a spot mask at a reciprocal distance a fractional % larger than the *a*-plane distance. The inverse FFT will then filter out *a*-plane distances/ bond lengths in flat regions that are smaller than the graphene *a*-plane distance (compression) as well as *a*-plane bonds on inclined surfaces, whose projected bond lengths therefore appear shorter than those of the graphitic *a*-planes in 'flat projection'. The colour coding (temperature scale) in figure 1c reveals a ripple crossing the image area diagonally, as indicated by the overlaid cartoon in fig. 1b. The band structure of rippled graphene was then calculated by tight binding (colours in the cartoon indicate extended/tensile (red) or compressed (blue) bonds). The bandstructure at the Dirac points in fig. 1e predicts a bandgap of 0.19 eV at the K point of the Brillouin zone in the regions of tensile strain. Regions of compressive or low (to zero) strain (green) in the cartoon do not exhibit a bandgap. However for compressive strain (fig. 1d) the Dirac point lies below the Fermi energy by 0.074eV.

(ii) Nano-sculpting. Following previous observations by the authors of hole formation after metal-evaporation on graphene, via metal-mediated etching in the presence of oxygen [16], and decoration of these holes with metal atoms [17,18] (see fig. 2a), we here go on studying the effects of metal-decorated holes on excitations in the UV/vis for the case of Pd. Rather than evenly covering the graphene, Pd (as all metals) forms nanoparticles of few nm in diameter (fig. 2d, dark-dot clusters with yellow sheen). In line with previous studies of the metal-graphene interaction [17], all nanoparticles sit on contamination and do not reside on pristine graphene surfaces; neither have single Pd-atoms been detected on clean graphene patches. In explanation of this, theoretical works have predicted very low binding energies for metals on monolayer graphene [19]. Figure 2c shows an EFTEM image of an excitation energy window at 3-4 eV, extracted from a data cube acquired from 17.5 eV to -2.5 eV in reverse mode, from high to low energy, to avoid afterglow effects from the high-intensity low loss region with an energy selecting slit width of 1 eV, an energy step size of 0.5 eV and an exposure time of the extracted slice of 5 seconds. Enhancement of the excitation at ~3 eV can be seen around two holes in the graphene, which are decorated by individual Pd-atoms, seen as dark dots in fig. 2d). The very bright yellow regions (the intensity is displayed on the temperature scale) arise from surface plasmons in Pd nanoparticles. In order to obtain the energy filtered image overlaid in the middle of fig. 2c (orange/blue colours) the excitation in the π-plasmon region at 4.5 eV was assimilated by a Gaussian peak, providing a simple and very good fit to the experimental data, and subtracted from the data cube (from which so far only the ZLP had been subtracted; blue/yellow colours in large underlying image 2c). The enhancement is spatially confined to less than 0.5 nm along the edge of the hole. Line profiles of the integrated loss intensity in the 3-4 V energy regime across edges of holes (one such scan location is shown as black arrow in fig. 2c together with the corresponding intensity profile) show a relative enhancement over that of pristine graphene of up to 50%. Example spectra extracted from EFTEM cubes together with arrows pointing to the respective locations, are shown in fig. 2e. The excitation at ~3 eV is clearly largest at the hole's edge. It has been disputed whether excitations at these energies are plasmons. The criterion for a Drude plasmon is a negative-to-positive zero-crossing of the real part of the dielectric function at the plasmon

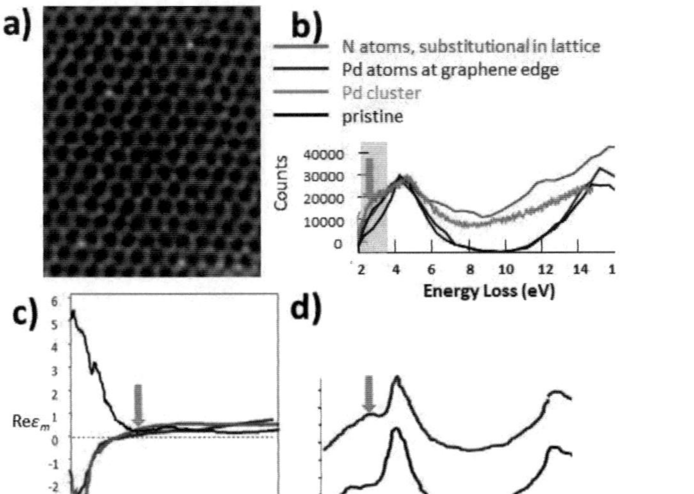

Fig. 3 a) Atomic resolution HAADF image (Nion UltraSTEM) of N-implanted graphene showing substitutional N atoms as bright dots (e.g., red arrow); **b)** experimental low-loss spectra extracted from EFTEM data cubes (TitanPICO); **c)** real part of the dielectric function extracted via Kramers-Kroenig analysis from the spectra in (b), of N-implanted, pristine and Pd-decorated hole-edges in graphene; **d)** simulated in-plane EEL spectra of a single Pd adatom on graphene in a 4x4 supercell, and of pristine graphene. The orange arrows point to the new feature at 3-3.5 eV in the experimental as well as the modelled spectra; it also denotes the zero-crossing in the dielectric function, indicating the plasmonic nature of this feature.

energy. A zero-crossing for spectra exhibiting the 3-eV feature can be seen in fig. 3c; it reveals that this feature is of plasmonic nature.

(iii) *Controlled impurity incorporation*. We have previously demonstrated that, in contrast to chemical doping or dosing via physi-sorption, it is possible to directly integrate metals, N and B into the lattice of graphene and CNTs via low-energy ion implantation [6,7] (see fig. 3a, and figs 4a&c). Figure 3b shows that, similar to Pd-dosing, N-doping introduces an excitation at ~3 eV in graphene, which is of plasmonic nature, as revealed by the zero-crossing of the dielectric function (fig. 3c).

Ion implantation of CNTs with Ag results in intercalation (fig. 4a) and a strong excitation at 3eV (fig. 4b). Out of the alkali metals, only K showed clear results, namely in in-plane doping (fig. 4c) and a plasmonic excitation at ~3 eV (fig. 4c, d & e). B doping also leads to lattice integration in both,

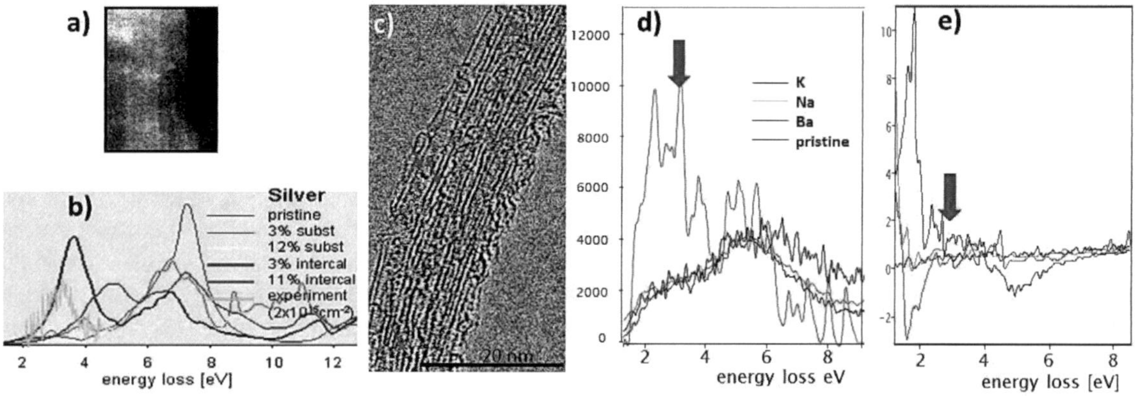

Fig. 4 a) Atomic resolution HAADF) image (Nion UltraSTEM) of an Ag-implanted MWCNT at 200 eV to a dose of $2x10^{15}$ cm^{-2}, revealing intercalated Ag-atoms (purple arrow); **b)** calculated (DFT: WIEN2K) low loss EEL spectra of MWCNTs with Ag-dopants in substitutional and intercalated positions, as well as experimental loss EEL spectrum of the tube in (a); **c)**) HREM image (TitanPICO) of K-doped DWNT bundle; dark dots (some are arrowed) are K-atoms integrated into the graphene sheets; **d)** low loss EEL spectra of DWNTs and SWNT bundles implanted with alkali and earth-alkali metals (TitanPICO). The spectrum intensity is normalised to the feature at ~5 eV (pi-plasmon). Doping with the metals was achieved by ion implantation at 200 eV to a dose of 10^{15} cm^{-2}; **e)** Reε_m of the dielectric function extracted via Kraemers-Kroenig analyisis from the experimental spectra (smoothed with a low pass filter with w= 0.2 eV).

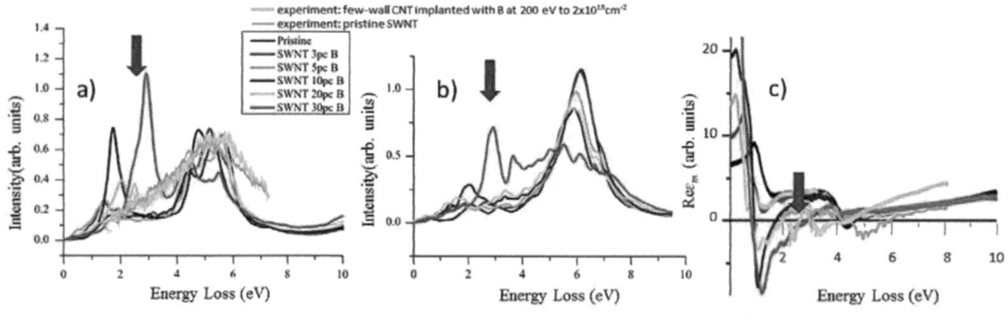

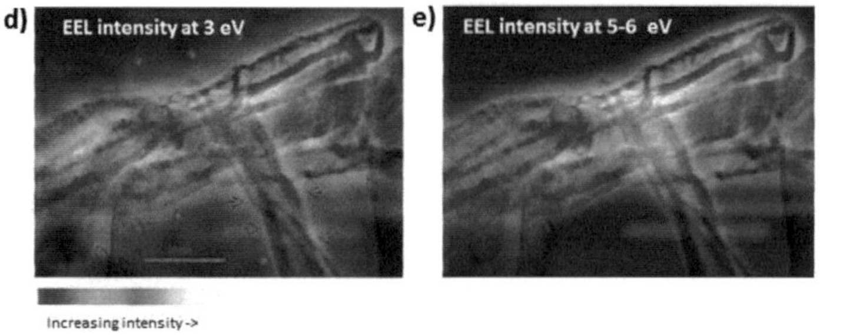

Fig. 5 a) Calculated electron energy loss spectra (DFT: WIEN2K) of a (substitutionally) boron doped (10,10) SWNT bundle in the out-of-plane and **b)** the in-plane direction, **c)** calculated out-of-plane component of the real part of the dielectric function. Also overlaid in (a) and (c) are experimental spectra of a heavily B-doped and a pristine carbon nanotube; **d)** background subtracted EEL map of an aggregate of double- and few-layer CNTs at 2-3.5eV and **e)** at 4.7-6.5 eV, extracted from an SI (NionUltraSTEM). The CNTs were ion-implanted at an energy of 200 eV, to a dose of $2x10^{15}$ cm^{-2}. The EEL map is overlaid on the Bright Field STEM image of the tubes (black shapes). Arrows in the left-hand image mark positions that indicate that the lower energy signal is less localised than the π-bulk plasmon (right) and more concentrated on the surface of the tubes; relatively higher intensity is seen here in the vacuum.

4. Conclusion. We have demonstrated the existence of ripples in graphene, which, according to calculations could shift the Dirac point and introduce a bandgap. Metal deposition on graphene leads to formation of metal ad-atoms decorating the holes, with creation/enhancement of a plasmonic feature at UV/vis energies along their edges. A similar feature is introduced by ion-implantation of N into graphene with N on substitutional sites, and of K and B into carbon nanotubes, with both species integrated into the carbon lattice. Ion implantation of Ag leads to intercalation, resulting in a similar excitation. It appears that heavier metals intercalate, whereas K, N and B integrate into the carbon lattice. All these impurities create a plasmon feature at around 3 eV.

Acknowledgements. We thank R. Gwilliam at the University of Surrey and J. Van den Berg at the University of Salford as well as their teams for ion-implantations into CNTs, and H. Hofsaess and J. Amani from the Universitaet Goettingen for ion implantations into graphene and CNTs. We are also indebted to R. Zan for preparation of graphene TEM samples at the University of Manchester and to R. J. Kashtiban from Warwick University for providing HREM images of graphene.

References.

[1] Banhart F, Kotakoski J and Krasheninnikov A V 2011 *ACS Nano* **5** 26
[2] Meyer J C , Kisielowski C, Erni R, Rossell M D, Crommie M F and Zettl A 2008 *Nano Lett.* **8 (11)** 2582
[3] Gass M H, Bangert U, Bleloch A L, Wang P, Nair R L R and Geim A K 2008 *Nat. Nano.* **3** 676
[4] Wang H, Wang Q, Cheng Y, Li K, Yingbang Y, Zhang Q, Dong C, Wang P, Schwingenschlögl U, Yang W and Zhang X X 2012 *Nano Lett.* **12** 141
[5] Bangert U and Zan R 2015 *Int. Mat. Rev.* **60(3)** 133
[6] Bangert U, Pierce W, Kepaptsoglou D M, Ramasse Q M, Zan R, Gass M H, Van den Berg J A, Boothroyd C B, Amani J and Hofsäss H C 2013 *Nano Lett.* **13 (10)** 4902
[7] Bangert U, Bleloch A, Gass M H, Seepujak A and Van den Berg J A 2010 *Phys Rev B* **81** 245423
[8] Li X, Cai W, An J, Kim S, Nah J, Yang D, Piner R, Velamakanni A, Jung I, Tutuc E, Banerjee S K, Colombo L and Ruoff R S 2009 *Science* **324** 1312
[9] Reina A, Son H, Jiao L, Fan B, Dresselhaus M S, Liu Z and Kong J, 2008 *J. Phys. Chem. C* **112** 17741
[10] XuY, Zhang K, Brusewitz C, Wu X and Hofsäss H C 2013 *AIP Advances* **3** 072120
[11] Hofsäss H, Ronning C and Feldermann H 2001 *Proc. 16th Int Conf on the Application of Accelerators in Research and Industry, Denton, TX, USA*, Nov 1-4 2000, eds. Jerome L Duggan and I Lon Morgan *AIP Conf. Proc.* **576** 947
[12] Eckstein W, Dohmen R, Mutzke A and Schneider R "SDTrimSP: A montecarlo code for calculating collision phenomena in randomized targets" 2007
[13] Blaha P, Schwarz K, Madsen G K H, Kvasnicka D and Luitz J 2001 *Wien2k User's Guide* http://www.wien2k.at/reguser/textbooks
[14] Bangert U, Gass M H, Bleloch A L, Nair R R and Geim A K 2009 *phys stat sol a* doi:10.1002/pssa.200824453
[15] Zan R, Muryn C A, Bangert U, Mattocks P, Wincott P,Vaughan D, Ruoff R S, Hamilton B and Novoselov K S 2012 *Nanoscale* **4** 3065
[16] Ramasse Q M, Zan R, Bangert U, Boukhvalov D W, Son Y-W and Novoselov K S 2012 *ACS Nano* **6 (5)** 4063
[17] Zan R, Bangert U, Ramasse Q M, Novoselov and K S 2011 *Nano Letters* **11(3)** 1087
[18] Zan R, Bangert U, Ramasse Q M and Novoselov K S 2012 invited Perspective *J. of Phys.Chem. Lett.* **3 (7)** 953
[19] Hardcastle T P, Seabourne C R, Zan R, Brydson R M D, Bangert U, Ramasse Q M, Novoselov K S and Scott A J 2013 *Phys. Rev. B* **87** 195430.

Nanodiamond stability with friction and heat

J Butterworth, K Briston, F Claeyssens and B J Inkson

[1]NanoLAB Centre, Department of Materials Science and Engineering, University of Sheffield, Sheffield, UK

E-mail: beverley.inkson@sheffield.ac.uk

Abstract. The structural stability of nanodiamonds (ND) has been investigated as a function of abrasion/friction and thermal annealing. Nanodiamonds examined before and after abrasion by transmission electron microscopy (TEM) are found to have modified surface characteristics. Frictional rubbing generates smaller, non-spherical particles with rougher, cleaner surfaces, and debris consisting of compacted ND+amorphous carbons. Annealing of ND and abraded ND at 1000°C demonstrates that abrasive pre-treatment increases proportion of transformation to onion-like-carbon (OLC) structures with generation of 3-30nm diameter OLC. The OLC achieved are of similar dimensions to those grown from ND annealed with a Ni catalyst but with more defective microstructures.

1. Introduction

Nanodiamonds (ND) can be fabricated by a number of methods including CVD, mechanical breakdown of bulk diamond and shock synthesis, with the latter method giving access to industrial scale manufacture [1]. Nanodiamonds have high hardness, strength and wear resistance, which together with their biocompatibility and distinctive optical properties gives rise to applications as polishing materials, lubricants, strong composites and potential as biomedical agents. Detonation nanodiamonds manufactured by shock synthesis are typically in the form of aggregations of ND nanoparticles, with cores of strongly bonded particles [1,2]. Aggregation, and mixed surface sp^n/sp^3 bonding, alters nanodiamond properties from bulk diamond.

Here the structural stability of detonation nanodiamonds has been investigated as a function of abrasion/friction and annealing at 1000°C. Detonation ND typically have their surface bonds terminated by graphitic or amorphous carbon and a range of stabilizing (C,H,O,N) functional groups [1,3]. These surface groups can allow the tailoring of nanodiamonds for a range of purposes such as drug delivery, but these stabilizing groups can also prevent graphitization and inhibit ND transformations [3]. The surface of diamond, and its termination by graphitic or amorphous carbon, may be modified by a range of methods including heat and mechanical shear [3-8]. Here the modification of ND by frictional abrasion has been investigated. Samples of ND are characterised by TEM before and after abrasion, and are found to have modified surface characteristics. Thermal annealing at 1000°C has been carried out to investigate the effect of abrasion on the ND-to-onion-like-carbon (OLC) transformation [4-6].

2. Abrasive stability of nanodiamond

Detonation nanodiamonds (NanoDiamond Products, Ireland) consisting of ~10nm diameter primary particles agglomerated in clusters (figure 1) have been subjected to abrasive wear tests and annealing, followed by TEM characterisation. ND in distilled water were dispersed ultrasonically before 50μl being dispersed onto a clean alumina sheet and dried. Abrasive testing between alumina counterfaces was carried out for 30 and 60min using a Buehler Minimet rotary polisher, 50rpm, F=98N. Abrasive debris was retrieved from the upper alumina platten by ultrasonic agitation in distilled water, and samples dispersed onto holey carbon grids for TEM (JEOL 2010F).

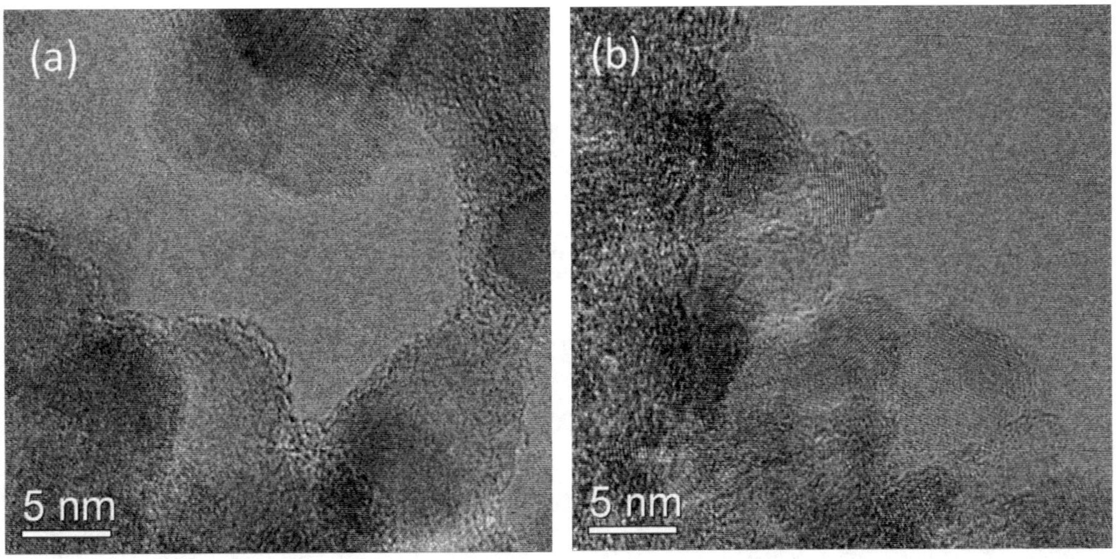

Figure 1. TEM images of nanodiamond particles. (a) As-received nanodiamond cluster with 5-10nm primary diamond particles enveloped in ~1nm of amorphous hydrocarbons. (b) 30 min abraded ND cluster.

As-received crystalline ND particles were typically 5-10nm in diameter, enveloped in ~1nm of amorphous hydrocarbons (figure 1a), and aggregated as clusters typical of detonation diamond [1-3]. The abrasive testing causes impacts between the ND clusters, and between ND clusters and the alumina substrate. Examination of the ND clusters after abrasion reveals that many have a significant reduction in the amorphous carbon layer observed on the as-received material. This is consistent with abrasion of the softer outer hydrocarbons and functional groups against other diamonds and the alumina plattens. It is also likely that there is local frictional heating during abrasion which may lead to loss of more volatile components. Frictional rubbing also generates smaller, non-spherical ND particles <5nm with rougher cleaner surfaces (figure 1b), which can originate from diamond-on-diamond wear. Some amorphous wear debris was also observed compacted with nanodiamond clusters. No evidence of graphitic or onion-like carbons was observed under these experimental conditions.

3. Thermal stability of abraded ND

ND are known to transform to onion like carbon (OLC) with thermal annealing [4]. Here samples of as-received ND and abraded ND were annealed at a temperature of 1000°C for 3 hours in vacuum furnace. This condition was chosen to ensure that the ND to OLC transformation would be in an early stage with partial ND→OLC tranformation [4] to investigate whether abrasion affects the transformation rate. Additional as-received ND was also annealed mixed with Ni nanoparticles (20nm diameter), as Ni is a known catalyst for carbon graphitization transformations. The ND and Ni were mixed ultrasonically in a ND: Ni ratio of 75:25wt% before drying prior to the annealing treatment.

3.1. Thermal stability of nanodiamond

Figure 2 shows examples of the microstructure observed for as-received ND and abraded ND after vacuum annealing at 1000°C for 3 hours. The as-received ND demonstrated the initial stages of the diamond to OLC transition, with graphitic layers visible on the surfaces of clusters. In some cases closed concentric graphite shells could be observed completely enveloping small diamonds (figure 2b). The partial covering of ND in a few graphitic layers at 1000°C is consistent with the ordering of carbon atoms on the surfaces of individual ND once surface groups (C,H,N,O) have thermally decomposed [4].

The annealed abraded diamond exhibited a distinctly different microstructure to the purely annealed ND. In addition to the growth of graphite on the diamond particles (figures 2a,b), a distribution of onion-like carbon balls 3-30nm in diameter (figures 2c,d) were observed throughout the annealed material. Figure 2d illustrates the structure of one of the larger OLC balls, which exhibit concentric but defective graphitic layering. These OLC originate principally from the amorphous hydrocarbon wear debris generated by the wear of the original ND surfaces, but also from complete transformation of the smallest diamond wear fragments.

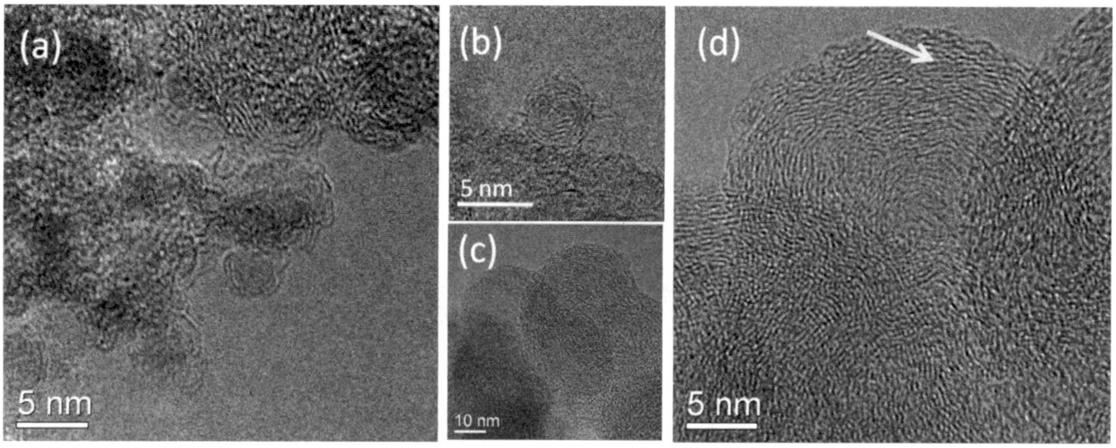

Figure 2. ND samples annealed at 1000°C for 3 hours. (a,b) As-received ND grows 1-3 layer graphite shells on diamond cores. (c,d) 30 min abraded ND annealing leads to growth of 3-30nm diameter OLC with defective graphite ordering (arrowed) from ND and amorphous carbon wear debris.

3.2. Thermal stability of nanodiamond+Ni nanoparticles

Figure 3 shows the microstructure observed for as-received ND annealed with Ni nanoparticles at 1000°C for 3 hours. The as-received ND annealed mixed with 25wt% Ni nanoparticles exhibited more conversion to graphite and OLC than purely annealed ND (figure 2a). ND particles were observed with well developed concentric shells of graphitic carbon, and graphitic sheets up to 10nm thick were also observed to envelop some Ni particles (figure 3a), indicating active diffusion of carbon and Ni throughout the sample. OLC of 2-25nm diameter were also observed, exhibiting very well ordered complete concentric graphite layers (figure 3b).

4. Discussion

Nanodiamonds have a range of potential uses exploiting the intrinsic properties of the diamond structure, however their high ratio of mixed sp^n/sp^3 surface bonding means that their activity is distinct from bulk diamond. Here abrasive wear of ND by collisions with other ND clusters and alumina counterfaces was observed to reduce the thickness of the surface hydrocarbons/functional groups, roughen the nanoparticle surfaces, and generate amorphous wear debris which compacted with the nanodiamond clusters. Under the abrasive conditions used, frictional shear/heating was insufficient to lead to graphitic debris, which has been observed *in situ* with single point contact shear of amorphous carbon [9].

Heating the detonation ND to 1000°C for 3 hours was sufficient to initiate the first stages of the diamond to graphite transformation, with onion-like shells of graphite growing around the ND particles [4]. The annealing of abraded ND however was quite different, revealing a highly accelerated transformation rate to OLC (figure 2c). Both small (<5nm) and large (>20nm) carbon onions grow, with large amounts of disorder in the concentric graphite shells. The abrasive process reduces the amorphous layer on the ND (mechanical displacement and frictional heating), causes wear of the ND surfaces (exposing active surface sites), and generates 3rd body amorphous wear debris; all of which enable enhanced carbon diffusion and growth of graphitic carbon during the subsequent annealing process.

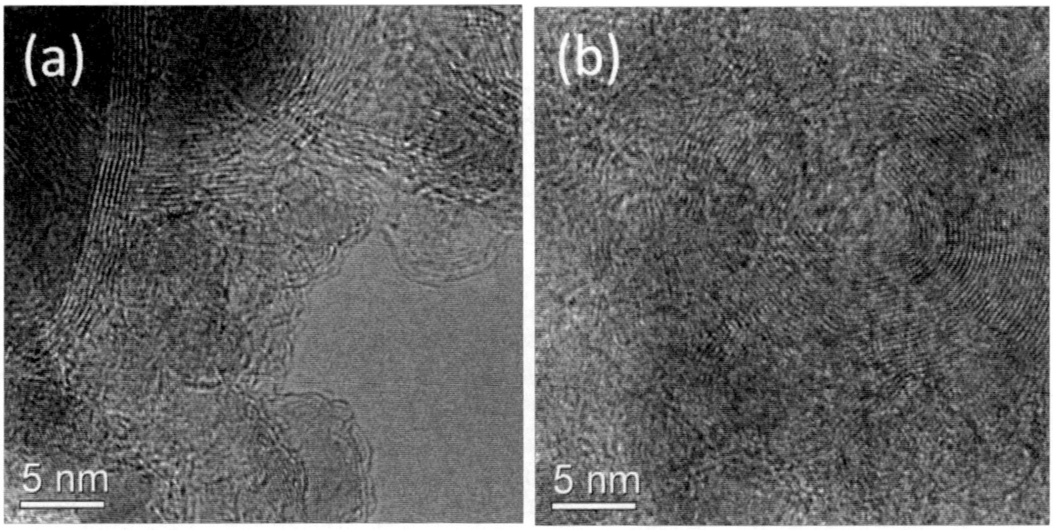

Figure 3. As-received ND+Ni nanoparticle sample annealed at 1000°C for 3 hours.
(a) Graphitization of the ND and Ni surfaces. (b) Growth of a highly ordered OLC sphere.

The role of friction and mechanical wear of ND clusters in accelerating the transformation rate of ND to OLC (figure 2), was compared to the use of catalyst Ni nanoparticles which also demonstrated increased transformation rate of ND to OLC (figure 3). Under the same thermal treatment (1000°C for 3 hours) OLC of similar dimensions were observed for both the abraded ND, and the ND+Ni mixture. The OLC generated from the abraded ND however had highly defective graphitic ordering, compared to the high-quality OLC with concentric graphite shells grown in the presence of diffusing Ni.

References

[1] Mochalin V N, Shenderova O, Ho D and Gogotsi Y 2012 *Nature Nanotech.* **7** 11
[2] Krüger A, Kataoka F, Ozawa M, Fujino T, Suzuki Y, Aleksenskii A E, Ya. Vul' A and Ōsawa E 2005 *Carbon* **43** 8 1722
[3] Kuznetsov V L and Butenko Y V 2005 *Synthesis, Properties and Applications of Ultrananocrystalline Diamond*, ed D M Gruen *et al* (Netherlands: Springer Press) p 199
[4] Kuznetsov V L, Chuvilin A L, Butenko Y V, Mal'kov I Y and Titov V M 1994 *Chem. Phys. Lett.* **222** 4 343
[5] Aleksenskii A E, Baidakova M V, Vul A, Davydov V Y and Pevtsova Y A 1997 *Phys. Solid State* **39** 1007
[6] Chen J, Deng S Z, Chen J, Yu Z X and Xu N S 1999 *Appl. Phys. Lett.* **74** 3651
[7] Butenko Y V, Kuznetsov V L, Churilin A L, Stankus S V, Khaitulin R A and Sagall B 2000 *J. Appl. Phys.* **88** 4380
[8] Pastewka L, Moser S, Gumbsch P and Moseler M 2011 *Nature Materials* **10** 34
[9] Wang J J, Lockwood A J, Peng Y, Xu X, Bobji M S and Inkson B J 2009 *Nanotechnology* **20** 305703

Quantification of airborne road-side pollution carbon nanoparticles

T Baquero[1], S Shukrallah[1], R Karolia[1], O Osammor[2], and B J Inkson[1]

[1]NanoLAB Centre, Department of Materials Science and Engineering, University of Sheffield, Sheffield, UK
[2]Air Quality, Monitoring & Modelling, Transport, Traffic & and Parking Services Division, Sheffield City Council, Sheffield, UK

Email: beverley.inkson@sheffield.ac.uk

Abstract. Roadside diesel particulate matter (DPM) has been collected using a P-Trak particle counter with modified inlet filter. The P-Trak monitor assesses ultrafine particle number in real-time rather than accumulated PM mass over a period of time, which is important for DPM where the particles are often <100nm in size. Collected pollution particulate matter was analysed by SEM and TEM, quantifying particle size, morphology and size distribution. The primary carbon nanoparticles form complex fractal aggregates with open porous morphologies and evidence of secondary carbon deposition. For the chosen collection sites, occasional but significantly larger mineral and fibrous particles were identified. The assessment of airborne particles by mass collection (TEOM), particle-number (P-Trak) and TEM methods is discussed.

1. Introduction

Airborne particles have potential to cause significant detrimental impact on human health, with the ability to penetrate deep into the human lungs depending on their size. Airborne particles come from a range of manmade and natural sources, with particular concern focusing on the range of $PM_{2.5}$ and $PM_{1.0}$ particulate matter that are generated by the burning of diesel fuels [1]. The World Health Organization (WHO) states that diesel emissions are carcinogenic to humans based on 'sufficient evidence that exposure is associated with an increased risk for lung cancer' [2]. The key outdoor air pollutants are ozone (O_3), nitrogen oxides (NO_x), particulate matter (PM) and sulphur dioxide (SO_2), causing 3.7 million premature deaths in 2012 [2].

There is clear evidence that airborne carbon-based diesel particulate matter (DPM) is concentrated in the vicinity of roads within our cities [3-5]. A variety of detection systems have been used to measure nanoparticle presence and exposure, including the scanning mobility particle sizer (SMPS) [6], aerosol mass spectrometer (AMS) [7], and ultrafine particle counter (UPC) [8]. Each method defines a different particle "equivalent diameter" based on the physical method of counting the number and sizes of the particles. The analysis of airborne particles by scanning and transmission electron microscopy (SEM, TEM) based methods [9, 10] is also important, to form a link between particle mass/number monitoring, and particle identity, since different particle chemistries and morphologies can exhibit different biological activity. The present study analyses the collection of airborne particles by UPC and evaluates particle size, morphology and size distributions using comparative SEM-based and TEM-based methods.

2. Collection of airborne road-side pollution

Airborne particles were collected using a TSI P-Trak Ultrafine Particle Counter (UPC) 8525. This instrument is based on the principle of condensation particle counting, able to detect particle concentrations up to $5x10^5$ particles/cm^3 with particle diameters from 20nm to 1μm at a constant total flow rate of $700cm^3$/min. Samples were collected concurrently with particle counting using TEM collection grids attached to the UPC inlet screen. An advantage of this method is that particles are collected directly onto the TEM grids without further manipulation.

Two holey carbon films on 300 Mesh Copper TEM Grids (AGS147, Agar Scientific) were welded to a coarse support mesh using silver conductive paint. Measurements were performed on April 29 2015 between 8:00AM and 9:30AM on the busy University of Sheffield roundabout where three major roads intersect. Sampling times were 15 minutes and 60 minutes, in order to evaluate time dependant particle effects. A tapered element oscillating microbalance (TEOM) teflon-coated glass fiber filter was used as reference to compare DPM agglomerates. The TEOM is located in the Sheffield City Council air pollution monitoring Station in Devonshire Green Park.

Transmission electron microscopy (JEOL JEM 3010) was used to evaluate size and morphology of the collected airborne aerosol particles. Scanning electron microscopy (FEI Inspect F) was used to evaluate particle distribution and characterise big agglomerates. The TEOM filter was gold-coated for 2 minutes at 15mA (Emscope, Quorum Technologies) to inhibit charging.

3. Characterisation of airborne pollution particulates

Figure 1 illustrates the types of particles collected in TEOM and UPC measurements. For both collection methods particles captured were dominated by carbon aggregates comprised of ~20nm diameter primary carbon nanoparticles (figure 1(b)). The macroscopic DPM agglomerates observed in the TEOM micrographs are formed due to gravitational sedimentation under controlled environmental conditions. Occasional larger particles were captured (arrowed figure 1(a)). Figure 2 illustrates some $PM_{2.5}$ and PM_{10} particles with complex shapes of both organic and man-made origin, that were observed decorated with smaller $PM_{1.0}$ particles.

Figure 1. (a) SEM image of airborne pollution accumulated on a TEOM filter. (b) TEM image of typical DPM particles collected by P-Trak UPC.

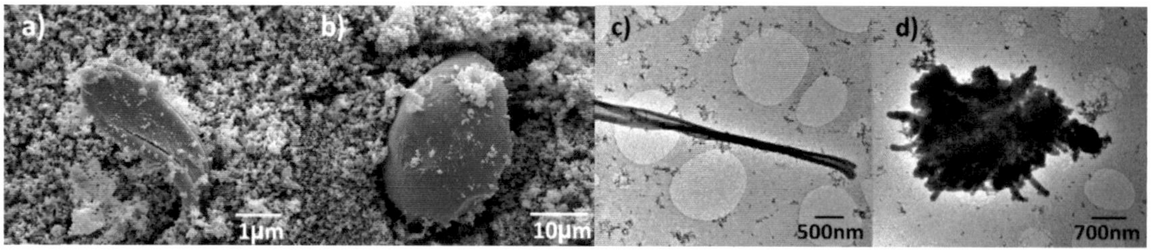

Figure 2. Larger $PM_{2.5}$ and PM_{10} pollution particles collected including (a,b) planar and ovalate particles, (c) nanoporous fibres, (d) fibre aggregates.

Individual DPM particles collected by P-Trak have a fractal-like morphology of agglomerated carbonaceous primary spherules [12]. In TEM projection the DPM aggregates appear well adhered to the holey carbon substrate (figure 3(a)), however their complex 3D fractal morphology means that they are frequently only attached at a few locations. The irregular 3D shapes can be characterised by calculating their radius of gyration taking into consideration the diameter and number of primary spherules of the aggregate [12].

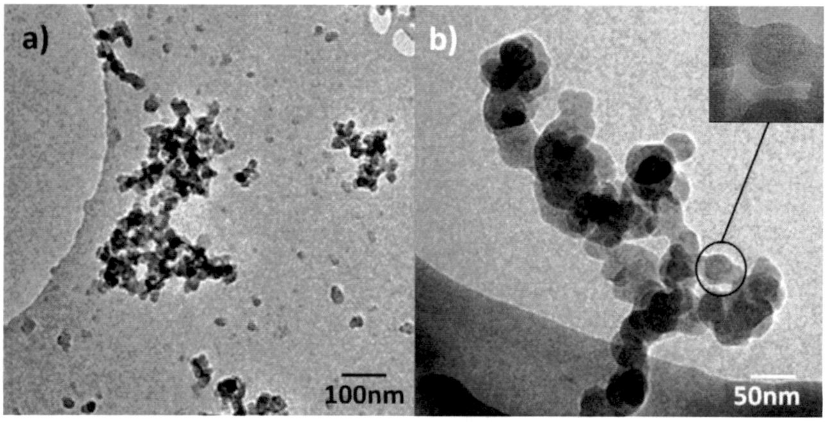

Figure 3. (a, b) Fractal morphologies of diesel soot particles comprised of nanoparticle aggregates with some evidence of secondary carbon layers (b).

To evaluate the number and size of DPM aggregates, the projected area of particles was measured by image analysis and an "equivalent diameter" calculated from a circle with the same area [11]. This method is accurate for spherical-like particles but does not accurately represent 3D fractal-like aggregates which can have very different projected cross-sections depending on imaging direction. Figure 4 shows the particle size distribution of 15 min and 60 min samples, analysed across the same total area of substrate by this methodology. It can be seen that more than 80% of the analysed particles were between the range of 18 and 24 nm which correspond to individual primary spherical particles. Larger aggregates represent less than 20% of particle number, but contribute a more significant percentage of particle mass [12].

Comparing the samples collected for 15 and 60 minutes, a marked variation in either of the projected density of particles or in their size distribution was not observed. This indicates that particles are not accumulating on the TEM grid with time, or merging as larger aggregates on the grid. This behaviour is consistent with a dynamical process whereby particles collide with the holey carbon TEM grid, a fraction adhering to it, but simultaneously a fraction of particles on the grid being dislodged in the incident airstream and being removed (figure 4(b)). This generates a steady state flow system where there is no net mass accumulation between 15 and 60 minute collection duration and a turbulent air flow as particles are observed in both sides of the grid.

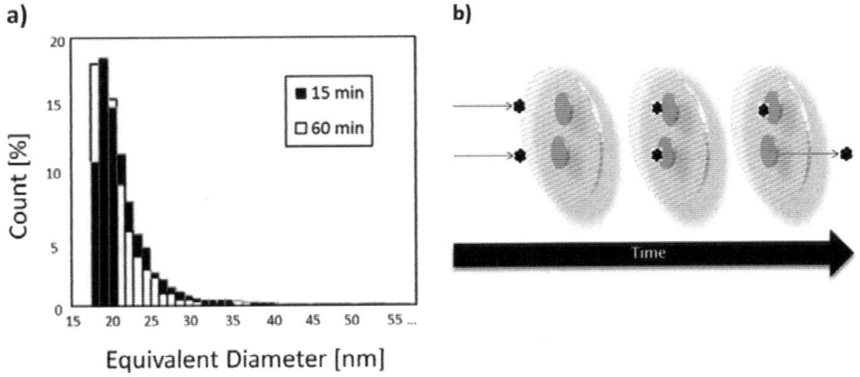

Figure 4. (a) Pollution particle size distribution of 15 min and 60 min samples collected by P-Trak. (b) Schematic of dynamical aerosol particle adhesion and removal on the TEM collection grids mounted on the P-Trak inlet screen

4. Conclusions

Airborne nanoparticles were collected directly into TEM grids using a commercial P-Trak UPC with modified inlet tube for further TEM and SEM analysis. Existing analysis methods assume DPM particles to be spherical [11] but they misrepresent larger soot aggregates which have fractal morphology. For 15min and 60min collection times on holey carbon TEM grids, there was no accumulation of DPM soot as a function of time. This is consistent with a steady state scenario whereby weakly bonded particles (limited contact area with the carbon film due to their fractal morphology) are constantly coming on and off the TEM grids. In the UPC the airflow may be turbulent as air passes through the holey carbon film and the coarse grid, and indeed particles were detected on both sides of the TEM collection grid. By comparison, in the TEOM method particles are collected in a slow gravitational airflow to assess mass accumulation. This study indicates the importance of local airflow to the dynamics of pollution particle adhesion to the collection media during aerosol sampling and assessment.

5. Acknowledgements

The authors gratefully acknowledge the HSL for equipment loan and discussions (Dr Gareth Evans and Dr. Delphine Bard).

6. References

[1] Burtscher H 2005 *J. Aerosol. Sci.* **36** 896.
[2] World Health Organisation. [Online]. Available from: http://www.who.int/mediacentre/factsheets/fs313/en/
[3] Nikolova I, Janssen S, Vos P, Vrancken K, Mishra V and Berghmans P 2011 *Sci. Total Environ.* **412** 336.
[4] Berghmans P, Bleux N, Panis I L, Mishra V and van Poppel M 2009 *Sci. Total Environ.***407** 1286-98.
[5] Cames M and Helmers E 2013 *Env. Sci. Eur.* **25**.
[6] Zhu Y, Hinds W C, Kim S and Sioutas C 2002 *J. Air Waste Manag. Assoc.* **52** 1032-42.
[7] Onasch T *et al* 2012 *Aerosol Sci. Technol.* **46** 804-17.
[8] Peters J *et al* 2014 *Atmos. Environ.* **92** 31-43.
[9] Murr L E and Soto K F 2005 *Mater. Charact.* **55** 50-65.
[10] Rmili B *et al* 2013 *Aerosol Sci. Technol.* **47** 767-75.
[11] Fujitani Y *et al* 2012 J. *Environ. Eng.* **S1**.
[12] DeCarlo P *et al* 2004 *Aerosol Sci. Technol.* **38** 1185-205.

Nanoscale investigation of organic – inorganic halide perovskites

S Cacovich[1], G Divitini[1], M Vrućinić[2], A Sadhanala[2], R H Friend[2], H Sirringhaus[2], F Deschler[2] and C Ducati[1]

[1] Department of Materials Science and Metallurgy, University of Cambridge, 27 Charles Babbage Road, Cambridge CB3 0FS, United Kingdom

[2] Cavendish Laboratory, University of Cambridge, JJ Thomson Avenue, Cambridge CB3 0HE, United Kingdom

Abstract. Over the last few years organic – inorganic halide perovskite-based solar cells have exhibited a rapid evolution, reaching certified power conversion efficiencies now surpassing 20%. Nevertheless the understanding of the optical and electronic properties of such systems on the nanoscale is still an open problem. In this work we investigate two model perovskite systems (based on iodine - $CH_3NH_3PbI_3$ and bromine - $CH_3NH_3PbBr_3$), analysing the local elemental composition and crystallinity and identifying chemical inhomogeneities.

1. Introduction
Mixed halide perovskites represent an emerging and widely investigated class of nanostructured materials. Recent studies have shown that the inhomogeneous properties of these materials are related to the grains distribution of the thin film [1] but a better knowledge of the local chemical composition is required to shed light on the still unknown processes underlying the local recombination centres in the films. In this context, the use of analytical electron microscopy could play a key role in the determination of elemental distribution at the nanoscale. The recent technical improvements, such as the introduction of high-brightness electron gun and large area EDX detectors, result in fast acquisition and limited electron dose on the sample, a crucial factor in the study of an organic – inorganic compound. Here we present EDX elemental maps of model perovskite systems acquired in STEM mode, and analysed using multivariate statistical analysis methods. In addition, High Resolution TEM was used to study the presence and the dimensions of crystalline domains within the films and their possible connections to the optoelectronic properties measured by optical methods.

2. Materials and methods
Methylammonium lead iodide ($CH_3NH_3PbI_3$) and methylammonium lead bromide ($CH_3NH_3PbBr_3$) were prepared according to standard procedure [2]. Bromine – based samples were synthesised following two different methods: using equimolar mixtures of methylammonium bromine (CH_3NH_3Br) and bromine iodide (BrI_2) and molar stoichiometric ratios of CH_3NH_3 and (Pb $(CH_3COO)_2$) in 20 wt% N,N – Dimethylformamide (DMF).
The perovskite films were analysed in a FEI Nova NanoSEM and in a FEI Tecnai Osiris TEM operated in scanning (STEM) and HRTEM mode at an acceleration voltage of 200 kV. The microscope is equipped with Bruker's Super-X EDX detectors and a high brightness electron gun (X-FEG), resulting in high count rates for elemental mapping. Elemental maps were acquired over areas

of 8x8 μm^2, with sampling every 100 nm and a dwell time of 100 ms per pixel, using a beam current of 700 pA and a probe size of 1 nm in diameter, and analysed using Hyperspy [3], an open source Python library dedicated to hyperspectral data analysis. Samples for TEM analysis were prepared by transferring scratched perovskite films on a TEM grid.

3. Results and discussion

Initially the morphology of the films was investigated. It was found that the precursor used in the synthesis process strongly affects the morphology of bromine film. $CH_3NH_3PbBr_3$ samples, prepared using $PbBr_2$ as precursor, present the typical perovskite crystallite structure, with cubes several microns in size, whereas the lead acetate based perovskite exhibits a much more homogenous film (**Figure** 1 a,b). The morphology of the iodine based film is completely different and the presence of elongated structures is evident in Figure 1 c.

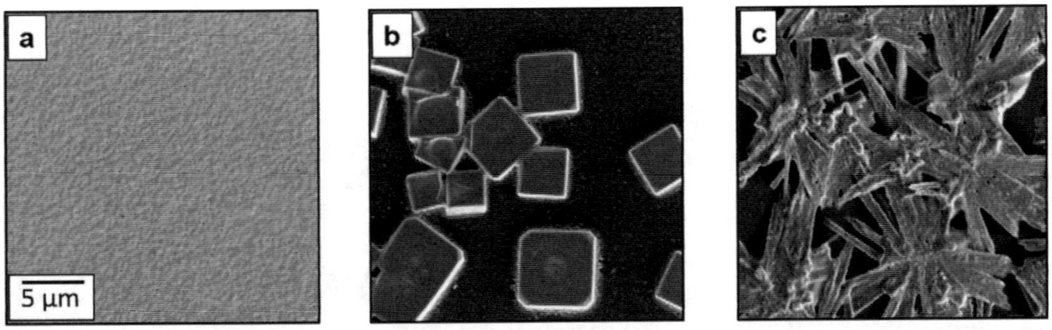

Figure 1. SEM images of a $CH_3NH_3PbBr_3$ synthesized using acetate (a), $CH_3NH_3PbBr_3$ (b) and $CH_3NH_3PbI_3$ (c). The scalebar applies to all images.

High Resolution TEM was used to study the crystallinity of the films to complement morphological information gathered from SEM analysis. The materials present crystalline domains, bigger in the bromine than in the iodine samples. HRTEM images (**Figure** 2) show a polycrystalline structure with adjacent domains of different crystal orientation on a length scale of 10s of nanometers. Fast Fourier Transforms (FFT) of the lattice show spacings compatible with the unit cell of known hybrid perovskite structures [4]. The materials do not appear amorphous. From these results it is possible to hypothesize a correlation between the local variation in photoluminescence [1,5] and phase and orientation of the perovskite domains, within both the bromide and the iodide based perovskites.

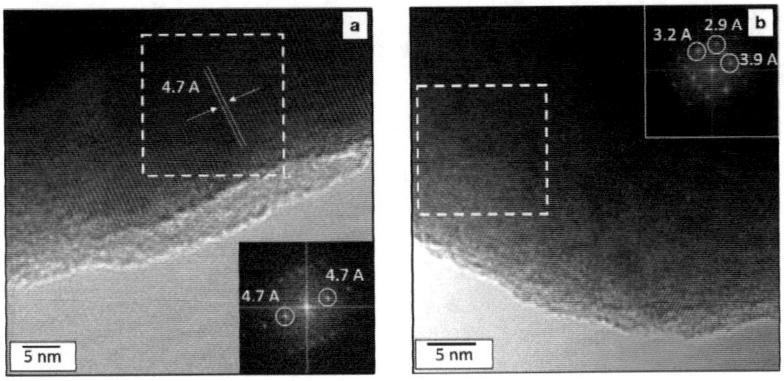

Figure 2. HRTEM images of the $CH_3NH_3PbBr_3$ (a) and $CH_3NH_3PbI_3$ (b) samples. The FFTs of the areas in the dashed squares are reported in the insets, with the visible lattice spacings highlighted. The $CH_3NH_3PbBr_3$ sample presents larger crystalline domains and the $CH_3NH_3PbI_3$ thin film is polycrystalline.

STEM EDX analysis was performed in order to study the local chemical composition and to assess the relative differences in the elemental composition. EDX data were treated using a PCA (Principal Component Analysis) – based routine. A model of the raw data has been reconstructed using a limited set of components, omitting those that contain only noise. Cliff Lorimer quantification method was applied on the denoised spectra, obtaining EDX elemental maps with nm-level spatial resolution. As shown in **Figure** 3 the differences between iodine and bromine EDX maps in both stoichiometry and homogeneity result quite clear. In the bromine based film the ratio Br/Pb is close to the nominal composition, whereas in the $CH_3NH_3PbI_3$ thin film the ratio is on average of the order of 1.2 and that is probably due to an incomplete conversion of the precursor in the synthesis process. The local variations in chemical composition are more evident in the pure iodine samples and I/Pb ratio varies between 0.92 and 1.36 in different regions. A detailed analysis to verify a direct link between photoelectronic properties and chemical variation is planned, and will be carried out in futher studies.

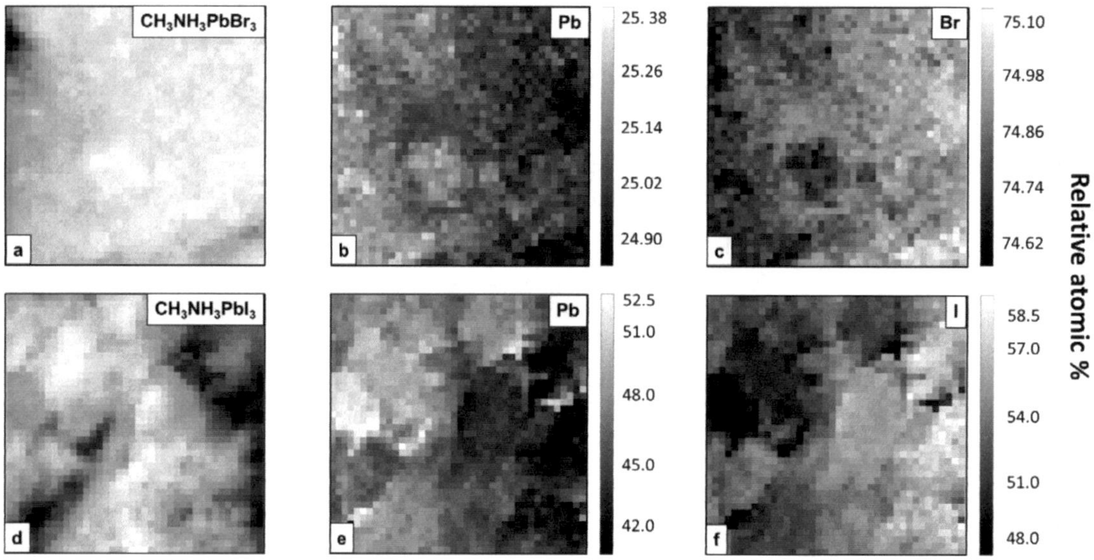

Figure 3. HAADF STEM images of $CH_3NH_3PbBr_3$ (a) and $CH_3NH_3PbI_3$ (d). Elemental maps of Br (b), Pb (c,f) and I (e) are reported for the two samples. All images are 4μm x 4μm.

Although STEM and TEM are techniques that provide crucial information on the nanostructure of the materials, the interaction of the electron beam with the sample can cause temporary or permanent changes to the samples. These phenomena are dependent on the atomic number of the elements in the specimen, their bonding, the electrical and thermal conductivity of the specimen. In a mixed organic-inorganic perovskite system, it is possible that different components of the crystal will be sensitive to beam damage in different ways, with the organic components of the crystal losing mass at higher rate than the inorganic Z-number component. Since precise quantification of light elements is problematic using EDX, and the main focus of this study was the compositional analysis of the inorganic component of the perovskite films, we only investigated the variations of Br, I, Pb (Cu signal from the support grid can also be seen in the spectra). Beam damage was tested acquiring EDX maps on the same area of the sample, using an acquisition time of 20s, 120s and 20s and a dwell time of 100ms per pixel. The analysis was carried out on a mixed halide perovskite film, with a halide mixture having a nominal ratio bromine/iodine:40/60. As shown in **Figure** 4 the chemical composition, after the acquisition of three spectra on the same area, does not change drastically.

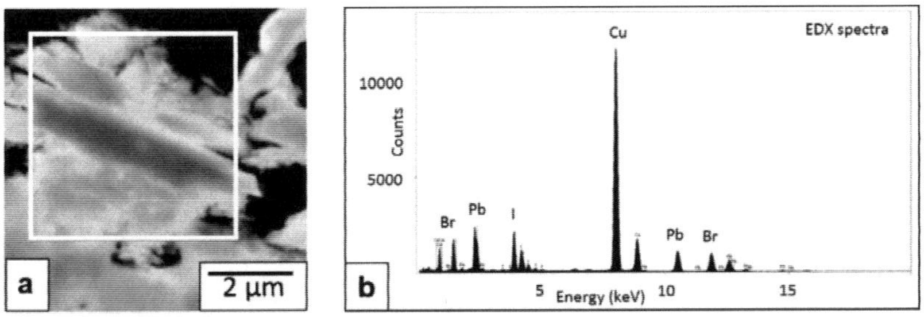

Figure 4. (a) STEM image of a mixed halide perovskite. The area in the white square has been analysed to test beam damage on the sample. (b) EDX spectrum of mixed halide perosvkite after 20s of acquisition time.

Table 1. EDX spectrum quantification of mixed halide perosvkite with an acquisition time of 20 s, 120 s and 2 0s. The quantification has been carried out using ES Vision

Element	Atomic % 20s	Atomic % 120s	Atomic % 20s
Cu(Kα)	76.72	76.01	75.55
Br (Kα)	08.82	09.64	10.01
I (Lα)	07.22	06.41	06.14
Pb (Lα)	07.22	07.93	08.16

4. Conclusions

Efficient elemental mapping in EDX-STEM allowed us to assess the chemical homogeneity of the inorganic component of Pb-Br and Pb-I perovskites, while high resolution TEM imaging demonstrated the nano-crystallinity of the materials. In particular, preliminary electron microscopy analysis shows that both orientation and phase of the crystallites and local elemental composition vary within the perovskite thin films.

In conclusion the use of STEM EDX analysis represents a powerful and versatile tool for the characterisation of devices based on hybrid composites with nano- and micro- scale structural and chemical features. Furthermore the latest developments in TEMs and data processing open up new opportunities for the investigation of optoelectronic devices, including beam sensitive materials, providing new information that can contribute to the understanding of the physical phenomena underlying the operation of solar cells, LEDs and lasers.

Acknowledgements

SC, GD and CD acknowledge funding from the ERC under grant number 259619 PHOTO EM. F.D. acknowledges funding from a Herchel Smith Research Fellowship. M.V. thanks the EPSRC, the Winton Programme for the Physics of Sustainability and the Cambridge Trust.

References

[1] de Quilettes D W, Vorpahl S M, Stranks S D, Nagaoka H, Eperon G E, Ziffer M E, Snaith H J and Ginger D S 2015 *Science* **348** (6235):683-6.

[2] Lee M M , Teuscher J, Miyasaka T, Murakami T N and Snaith H J 2012 *Science* **338** (6107) 643-7.

[3] de la Peña F, Burdet P, Sarahan M, Ostasevicius T, Taillon J, Eljarrat A, Mazzucco S, Fauske V T, Donval G, Zagonel L F, Iyengar I and Walls M 2015 Hyperspy 0.8 10.5281/zenodo.16850.

[4] Im J H, Chung J, Kim S J and Park N G 2012 *Nanoscale Research Letters* **7**:353

[5] Vrućinić M, Matthiesen C, Sadhanala A, Divitini G, Cacovich S, Dutton S E, Ducati C, Atatüre M, Snaith H J, Friend R H, Sirringhaus H and Deschler F 2015 *Preprint* advs.201500136.

Morphological and Compositional (S)TEM Analysis of Multiple Exciton Generation Solar Cells

F Wisnivesky-Rocca-Rivarola[1], N J L K Davis[2], M Bohm[2], C Ducati[1]

[1] Department of Materials Science and Metallurgy, University of Cambridge, 27 Charles Babbage Road Cambridge, CB3 0FS, UK,

[2] Cavendish Laboratory, University of Cambridge, J.J. Thomson Avenue, Cambridge, CB3 0HE, UK

E-mail: fw299@cam.ac.uk

Abstract. Quantum confinement of charge carriers in semiconductor nanocrystals produces optical and electronic properties that have the potential to enhance the power conversion efficiency of solar cells. One of these properties is the efficient formation of more than one electron-hole pair from a single absorbed photon, in a process called multiple exciton generation (MEG). In this work we studied the morphology of nanocrystal multilayers of PbSe treated with $CdCl_2$ using complementary imaging and spectroscopy techniques to characterise the chemical composition and morphology of full MEG devices made with PbSe nanorods (NRs). IN the scanning TEM (STEM), plan view images and chemical maps were obtained of the nanocrystal layers, which allowed for the analysis of crystal structure and orientation, as well as size distribution and aspect ratio. These results were complemented by cross-sectional images of full devices, which allowed accessing the structure of each layer that composes the device, including the nanorod packing in the active nanocrystal layer.

1. Introduction

Formation of multiple electron-hole pairs following absorption of a single energetic photon presents a possible route to improved photovoltaic conversion efficiency, since an additional photocurrent can be generated while bypassing heat generation resulting from electron-phonon scattering. The mechanism of converting excess photon energy in additional charge carriers can increase the efficiency limit of single junction photovoltaic devices to over 40% [1].

The extent to which a hot charge carrier undergoes the MEG process is determined by several factors, such as the strength of the Coulomb coupling between the hot charge and the trion (which consists of a charge carrier at lower energy, formed when a hole or electron is trapped on the QD surface, and an additional electron-hole pair), and by the density of final trion states coupled to the initial hot charge carrier state. The influence of these factors on the MEG process depends on the semiconductor nanocrystal shape and size [2]. In nanocrystals where the physical radius is of the order of, or smaller than, the exciton Bohr radius, quantum confinement leads to discrete excitonic energy levels, and the dimensions of the nanocrystals determine the spatial extent of excitons [3]. Therefore, morphological characterisation of the nanocrystals, as well as the full device architecture is essential to understand how the fundamental processes of these devices are being affected and identify ways to improve device efficiency.

In this work we analysed the morphology individual PbSe nanorods treated with $CdCl_2$ and full

MEG deviced using analytical electron microscopy to characterise their chemical composition and morphology. Lead chalcogenite nanocrystals are interesting for these type of devices due to their small band gap, which is small enough to allow MEG from low energy photons [3]. In addition, the relatively large dielectric constant of Pb chalcogenite materials results in a reduction of carrier-carrier interactions, leading to an increase in the exciton Bohr radius, which could potentially be of the order of the length of the nanorod [4]. The samples studied were probed by combining high-resolution transmission electron microscopy (HRTEM), scanning TEM (STEM) and energy dispersive X-ray spectroscopy (EDX) with focused ion beam (FIB) milling for preparation of cross-sectional specimens. This approach revealed the structure of the different solar cell components at length scales of 1-100 nm.

2. Experimental methods

PbSe nanorods (NRs) of three different bandgaps (1.05 eV, 0.95 eV and 0.80 eV) were synthesized following a method adapted from Koh et al [5], with an additional $CdCl_2$ treatment at the end of the NR synthesis to provide additional surface passivation. The relevant energy levels for the different NCs solutions (conduction band edge and valence band edge) were determined by using a combination of ultraviolet photoelectron spectroscopy (UPS) and absorbance spectroscopy, as described by Ehrler et al. [6].

Samples for electron microscopy analysis were prepared by drop-casting diluted NR solutions on TEM copper grids. High-resolution TEM (HRTEM) and high angle annular dark field (HAADF) STEM images were obtained using a FEI Tecnai F20 microscope operated at 200 kV. For the preparation of the TEM specimens for cross-sectional view, a full solar cell device was thinned by focused ion beam milling (FEI Helios dual beam FEG SEM/FIB microscope), with a final thickness of around 100nm. Energy dispersive X-ray spectroscopy (EDX) mapping of the cross-sectional specimen was performed using a FEI Tecnai Orisis TEM/STEM operating at 200 kV.

3. Results and discussion

Figure 1 shows HAADF-STEM images of PbSe nanorods presenting three different band gaps, namely: 0.80 eV, 0.95 eV and 1.05 eV. Low-magnification images indicate that PbSe nanocrystals display the expected rod-like morphology, and are monodisperse with uniform diameter and length. The 1.05 eV NRs displayed a less uniform morphology, and a fraction of the nanorods were not straight, exhibiting kinks and bends. The average diameter for the samples was of the order of 4-6nm and average length of around 10-16nm. Histograms presented in figures 1(a) and 1(b) show the size distributions for the nanorods of each band gap. Previous studies [7,8] correlated the aspect ratio of PbSe nanorods with the carrier multiplication and Auger recombination, indicating that carrier multiplication in NRs occurs in a similar way to that in QDs, but with an enhanced Coulomb interaction due to the one dimensional character of electronic states [9]. The enhanced Coulomb interaction increases the carrier multiplication efficiency in NRs with moderate aspect ratios of up to 6 [8]. The results shown in Figure 1(c) and 1(d) indicate that the aspect ratio of the samples analysed is around 2.5-3, being within the values considered favourable for an increase in carrier multiplication, as shown in work by Padilha et al. [8].

Detailed crystal structures of single PbSe nanorods were characterized by HRTEM, as shown in Figure 2. The analysis of different parts of each sample indicated that the nanorods are single-crystalline with a crystal structure is rocksalt and a growth along the [0 0 2] direction. STEM-EDX was used to obtain elemental maps of the PbSe nanorod films, as presented in Figure 1 (b)-(e). The maps obtained provide a qualitative evaluation of the nanorod film composition. The maps presented in Figure 1 confirm the composition of the nanorods and indicate the presence of the $CdCl_2$ passivating layer, which is important to reduce the density of trap states in the surface of the nanorods, as demonstrated by Ip et al. [10].

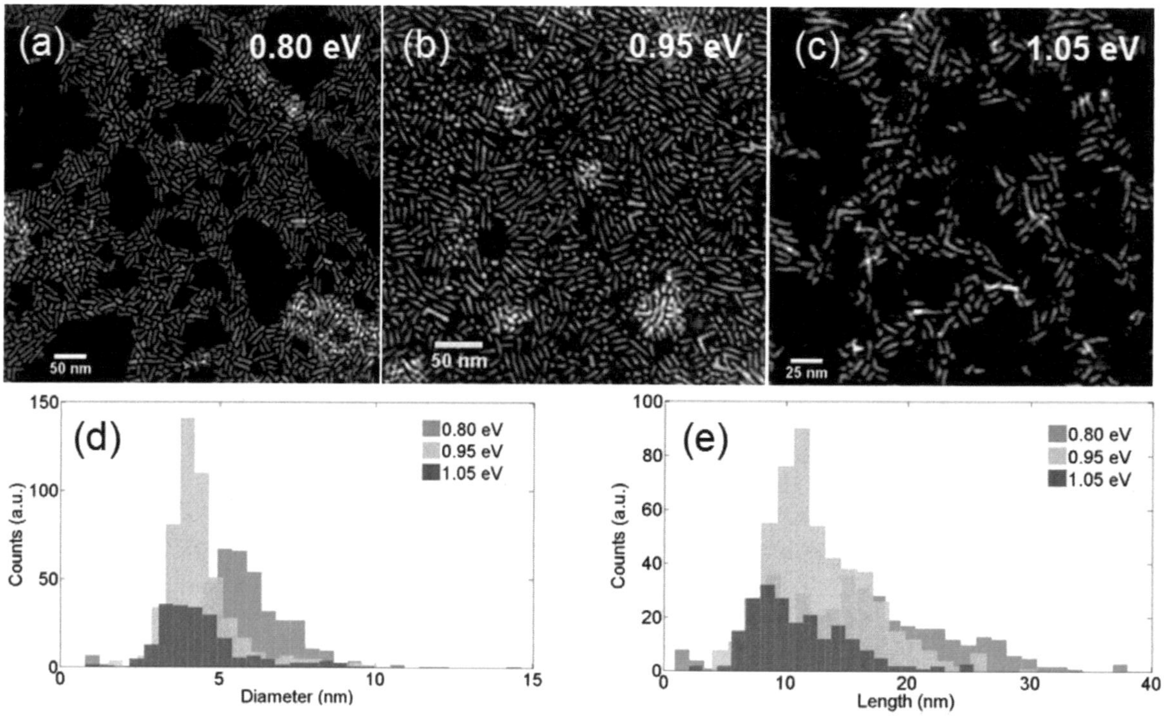

Figure 1. HAADF-STEM images of PbSe nanorods of the 3 different band gaps: (a) 0.08 eV, (b) 0.95 eV and (c) 1.05 eV. (d) and (e) show the histograms for the diameter and length distributions of the nanorods, respectively.

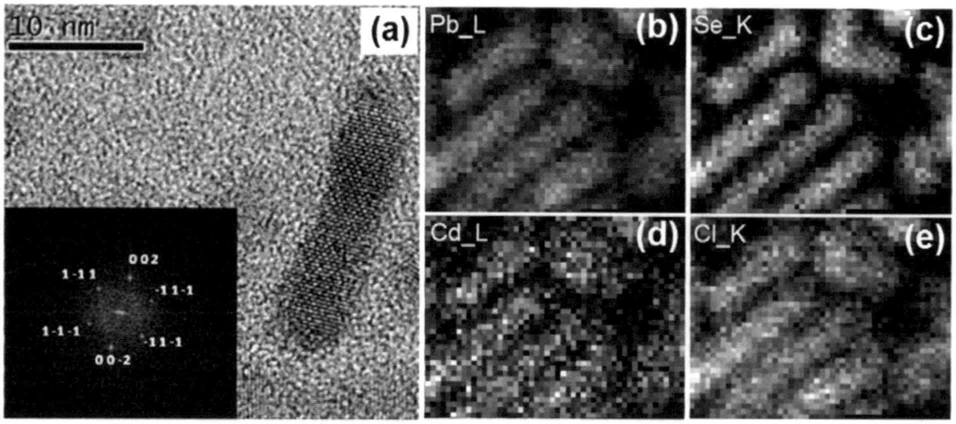

Figure 2. (a) HRTEM image of a PbSe nanowire. the inset shows the corresponding FFT pattern that shows that the nanorod is viewed on the [1 1 0] zone axis, (b)-(e) STEM-EDX elemental maps of the PbSe nanorods with a $CdCl_2$ passivation layer.

The nanorod solutions analised were used to produce full solar cell devices by depositing an array of PbSe NRs on a ZnO film which was produced using a sol-gel method. The NRs were deposited in a layer-by-layer approach using the ligand 1,2-ethaneditiol (EDT) for the first layers and hydrazine as the exchanging ligand for the final NR layer. The final device presented an efficiency of 2.5%, and a external quantum efficiency of 120%. Figure 3 shows a HAADF-STEM image of a cross-section of the full device. The micrograph shows a clear difference in NR packing between the layers deposited with EDT (right of the dashed line), which presents a thickness of 56 nm, and hydrazine (left of the dashed line) with a thickness of 48 nm. The hydrazine treatment appears to improve NR packing, reducing the distance between the rods and forming a uniform dense layer, with no cracks or pinholes, which should result in an enhanced film conductivity.

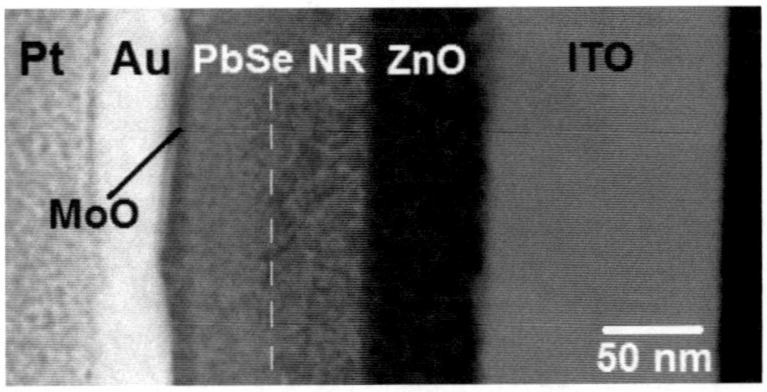

Figure 3. Cross-sectional TEM outlining the device architecture (rotated left 90°). The dashed line indicates the interface between the two ligands used for deposition of the NRs: EDT (right) and hydrazine (left).

4. Summary

Detailed morphological and compositional study of PbSe NR films showed that the nanocrystals presented a uniform size distribution with an aspect ratio of the order of 2.5-3, being within the values considered favourable for an increase in carrier multiplication in MEG devices. Additional STEM-EDX analysis gave qualitative information of the NR films composition, indicating the presence of the $CdCl_2$ passivating layer, responsible for improving surface conductivity in the NRs. Further analysis of cross-sectional specimens prepared from full devices using FIB technique revealed that the hydrazine treatment used during PbSe deposition improves NR packing, forming a uniform film.

References

[1] Nozik A J 2010 *Nano letters* **10** 2735
[2] Liu Z, Sun Y, Yuan J, Wei H, Huang X, Han L, Wang W, Wang H, Ma W 2013 *Advanced Materials* **25** 5772
[3] Cunningham P D,Boercker J E, Foos E E, Lumb M P, Smith A R, et al. 2011 *Nano Letters***11** 3476
[4] Sills A, Califano M 2015 *Phys. Chem. Chem. Phys* **17** 2573
[5] Koh W K, Bartnik A C, et al. 2010 *Journal of American Chemical Society* **132** 3909
[6] Ehrler B, Walker B J, Bohm M L, Wilson M W B, et al. 2012 *Nat. Commun.* **3** 1019
[7] Padilha L, Stewart J T, Sandberg R L, Bae W K et al. 2013 *Accounts of Chemical Research* **46** 1261
[8] Padilha L , Stewart J T, Sandberg R L, Bae W K, Koh W K et al. 2013 *Nano Letters* **13** 1092
[9] Bartnik A, Efros A L, Koh W K, Murray C B, Wise F W 2010 *Physical Review B* **82** 1
[10] Ip A H, Thon S M,Hoogland S,Voznyy O,Zhitomirsky D et al. 2012 *Nature nanotechnology* **9** 577

Acknowledgments

The authors would like to thank financial support from: ERC (grant number 259619 PHOTO-EM), EU (grant number 312483 ESTEEM2), and CNPq (grant number 246050/2012-8).

High Resolution Quantitative Lorentz Microscopy

S McVitie, D McGrouther and M Krajnak

Scottish Universities Physics Alliance, School of Physics and Astronomy, University of Glasgow, Glasgow G12 8QQ UK

stephen.mcvitie@glasgow.ac.uk

Abstract. The advent of aberration corrected transmission electron microscopy has led to considerable improvements in the field of high resolution electron microscopy imaging. In this paper we show how these developments are applied to imaging of magnetic structure in field free or low field conditions. Whilst the capability of increased spatial resolution is demonstrated on magnetic layers with a width of < 20nm we also consider how a pixelated detector can be used to dramatically increase the efficiency of the detection of the magnetic signal variation in the presence of strong diffraction contrast.

1. Introduction

The study of magnetic materials in the transmission electron microscope has a long and rich history dating back to the 1950s. Imaging of magnetic structure in the TEM comes under the umbrella of methods termed Lorentz microscopy [1]. Quite simply such imaging can be described in terms of the interaction of the electron beam with the magnetic induction through the classical Lorentz force. Lorentz microscopy as a form of phase imaging can also be considered from a quantum mechanical wave-optical approach, although the interpretation is slightly less intuitive. The latter approach is required for any image calculation and incorporation of microscope parameters. A range of methods exist within TEM such as the defocused Fresnel and displaced aperture Foucault which can be used to image domain walls and domains respectively, albeit these are non-linear methods where quantification can be problematic. In scanning TEM the differential phase contrast (DPC) technique provides quantitative induction maps by mapping the displacement of the electron beam on a segmented detector.

In principle Lorentz microscopy can be practiced on any TEM/STEM, although in order to image the magnetic structure the sample requires to be in field free or a low field environment. For the majority of specimens studied this therefore requires the objective lens to be switched off or be weakly excited. As the objective lens is not used for imaging one has to rely on lenses located further from the sample which are less powerful than the objective lens. This means that the resolution attainable in Lorentz microscopy is generally at least one order of magnitude less than that possible with high resolution TEM. The recent advent of commercially available aberration corrected instruments has transformed the capability of (S)TEM and this has been exploited successfully for high resolution imaging [2]. At Glasgow we have pioneered the use of aberration corrected Lorentz STEM with the DPC mode being used to demonstrate sub nanometer resolution in field free/low field conditions [3].

In this paper we highlight the capabilities that have been afforded with the system now installed at Glasgow. Whilst the superior resolution is one of the main factors of the aberration corrected system we have also explored possibilities of novel detection where the sensitivity of DPC to small deflection angles (phase gradients) has shown extremely promising results.

2. Experimental set up

As stated previously one of the major advances in (S)TEM imaging recently has been the incorporation of aberration correctors and the resulting increase in spatial resolution. At the University of Glasgow we have a JEOL Atomic Resolution Microscope (JEM-ARM200F) equipped with a CEOS (Corrected Electron Optical Systems GmbH) probe corrector for STEM imaging. This system has a cold field emission gun and operates at accelerating voltages of 60, 80 120 and 200 kV. With the objective lens normally excited the field at the sample is around 20000 Oe (2.0 T). As stated in the previous section Lorentz imaging is carried out in low field or field free conditions. This instrument is flexible and we can control the field at the sample from 0 Oe up to the full objective strength. The remanent field of the objective lens when it is switched off is 150 Oe and for many samples we image with this field present. If lower fields are required we have a control system designed at the University of Warwick which can reduce the field at the sample to 0 Oe. Furthermore we can set any field with the objective lens so that we can observe magnetisations processes by simply tilting the sample with the goniometer.

On this instrument we perform DPC imaging with an 8 segment silicon photodiode array detector (supplied by Deben UK Ltd.) comprising a solid quadrant inside a contiguous annular quadrant. Signals from the detector were amplified using the "Superfast" amplifier (Andrew Armit Designs). For full flexibility and real time display of magnetic induction information the 8 detected signals were acquired, mixed and displayed via four Gatan Digiscan II units which mean that up to 16 signal combinations may be displayed simultaneously.

We have also explored DPC imaging with pixelated detectors. Initial experiments were carried out using a CCD camera (Gatan Orius) to record the bright field disk. A CCD camera is not ideally suited for STEM imaging due to issues such as charge spreading and its low frame rate. However here it was used to demonstrate the principle of improving the DPC imaging by recording the diffraction disk at each point in the scan. Subsequently using image processing on this data set allowed an analysis of the beam deflection. The Orius camera has high resolution (11MPix resolution, 4008×2672 pixels) but was binned 4× to reduce the large amount of data associated generated by recording a diffraction images at each point in the scan. The timescale of such imaging was also problematic due to the maximum frame rate of the camera being only 14 Hz. For the imaging carried out here we used a beam dwell time of 50 ms to obtain a 100×100 pixel STEM image, further taking account of the camera readout time the images were acquired over 12 minutes. Realistically such long acquisition times are not practical as both stability and sample drift issues are of considerable concern, particularly if larger image sets are to be taken. To overcome this a direct solid state detector is being investigated which will allow shorter acquisition times. With this in mind we have are looking at a direct radiation pixelated detector known as Medipix3 [4]. This detector is currently being tested and we expect this will result in acquisition times being reduced by at least an order of magnitude.

3. Imaging considerations

DPC imaging has been used to image many magnetic systems producing detailed induction maps to aid our understanding of nanomagnetic structures. However it should be noted that the Lorentz interaction for many thin films results in a very small deflection of the electron beam. As an example consider a uniformly magnetized thin film as shown in Fig. 1(a), the interaction of the electron beam with the thin film results in a magnetic deflection angle, β_L, which can be shown to equal $(e\lambda/h)B_St$ where B_S is the saturation induction of the material, t is the film thickness. The fundamental constants are the magnitude of electronic charge (e), electron wavelength (λ) and Planck's constant (h). For reference a typical value for a permalloy ($Ni_{80}Fe_{20}$) film with B_S = 1.0 T and thickness 10 nm we calculate β_L to be 6.4 μrad. This compares with typical Bragg angles which are 10 mrad or more. In terms of DPC STEM imaging, a schematic of the arrangement is shown in Fig 1(b) where the local deflection of the beam is detected via a quadrant detector. By taking difference signals from opposite segments of the detector components of the integrated induction can be mapped. A point to note is that only the bright field disk is detected on the quadrant detector.

When the objective lens is switched off in the JEOL-ARM200F the probe angles subtended at the sample are much lower than for HREM being in the range of a few mrad and below. With the probe aberration corrector the optimum probe angle (α) is 3.2 mrad which give a diffraction limited probe size of around 1.5 nm and a resolution half of this value [3]. Whilst this is very high resolution for magnetic imaging it is also important to consider the sensitivity of the technique in being able to measure small deflections. We can consider this in more detail by looking at the lengthscales (spatial frequencies in the detector plane). This is illustrated in Fig 1(c) for the case of an annular split detector comprising two segments, with an electron beam being displaced due to a Lorentz force or phase gradient. In the detector plane we can consider angular quantities or spatial frequencies which we can equate, e.g. $k_\alpha = \alpha/\lambda$. For an optimum probe angle of 3.2 mrad and a 200 kV beam with $\lambda = 2.51$ pm, we have a spatial frequency $k_\alpha = (0.8 \text{ nm})^{-1}$. The important spatial frequencies shown in the figure are the probe radius (k_α), the inner radius of the annulus (k_i) and the spatial frequency associated with the beam deflection (k_β). We should note also that for the typical Lorentz deflection angle given earlier (6.4 μrad) this corresponds to a very low spatial frequency $k_\beta = (400 \text{ nm})^{-1}$, note that the deflection in the figure is exaggerated. The annular detector is the basis of modified DPC in which it was shown to provide a suppression of the non magnetic signal variation due to differential Bragg scattering from the 5-10 nm diameter crystallites comprising the thin film compared to standard DPC [5]. This filtering of the high spatial frequency signal relative to the low spatial frequency magnetic signal variation becomes more enhanced as the ratio (k_i/k_α) becomes close to unity. It is evident that the spatial frequency associated with the magnetic films is considerably smaller than bright field disk by almost two orders of magnitude. We will discuss this further in the next section. Recently the use of pixelated detectors to record STEM diffraction patterns has allowed a number of advances. These have resulted in making further efficiencies in enhancing the phase information in STEM DPC imaging [6].

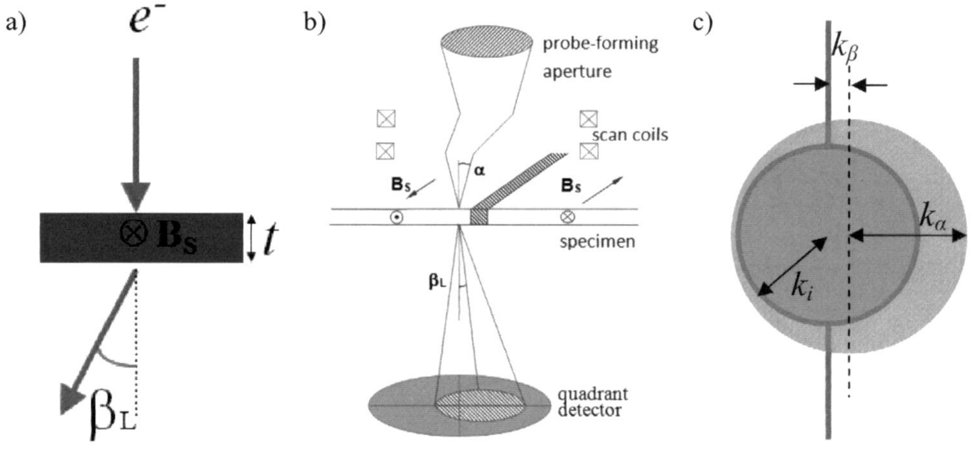

Fig 1. a) Schematic of the deflection of the electron beam from a uniformly magnetised (saturation induction B_s) film of thickness t. b) STEM DPC imaging schematic showing the key quantities of probe angle, α, and the Lorentz deflection, β_L. c) Schematic of bright field disk falling on annular split detector, showing the key quantities in reciprocal space: the probe radius, k_α, magnetic spatial frequency, k_β, and inner detector radius, k_i.

In the following section we demonstrate the resolution improvement capability that has been achieved with the STEM corrector system with an example from a magnetic thin film system possessing nanometer scale magnetic structure. Furthermore we look at the possibilities of improvements for magnetic imaging with pixelated detectors.

4. Results

In this section we present results which demonstrate the possibilities with the aberration correction system. Firstly the capability in terms of improved spatial resolution is considered, before moving on to looking at how pixelated detection opens up potentially huge improvements in DPC imaging.

We illustrate the power of the resolution of the aberration corrected system on an exchange biased sputtered multilayer sample [7]. This sample comprises a structure of ferromagnetic NiFe (permalloy) layers between anti-ferromagnetic FeMn layers with the structure NiFe/(FeMn/NiFe)×10 grown on an oxidised Si substrate with a capping layer of 5nm of Ta. The NiFe layers have an average thickness of 16.5 nm and the FeMn layers an average of 12.8 nm, measured from high resolution STEM imaging. Such a structure is too thick for plan view TEM even if the substrate was removed, so for the experiments conducted here a cross-section sample was prepared using focused ion beam methods resulting in a section of ~ 80 nm thickness. Measurements on the initial exchange bias coupling between the AF and FM layers shows hysteretic behaviour in the continuous film sample with the individual layers reversing in steps so that the sample reversed completely in a field of ~150 Oe. For the cross-sectioned sample however the magnetic configuration has an additional factor - in effect the cross-sectioning has patterned each FM layer into a nanowire geometry and the fields required to reverse each layer are consequently much higher. Using the JEOL ARM we easily imaged the state of each FM layer, which had a lateral width of < 20nm. Furthermore using the objective lens field we were able to reverse the moment in the layers by tilting the sample with an in-situ field. An example of the configuration of the cross-section partway through the reversal is shown in the DPC image in Fig. 2(a). Here the component of magnetic induction mapped is parallel to the layer interfaces and it can be clearly seen by the black/white contrast that the FM layers are pointing in opposite directions. Furthermore the antiferromagnetic FeMn layers are grey indicating, as expected, no net induction component in these regions. Whilst the magnetic state of each of the 11 layers can be clearly determined to be up or down in the image, there is a significant variation in contrast within each stripe. For such a layered structure sectioned here it would be expected that the magnetisation should be uniform, whilst the DPC contrast suggests it is not. However the contrast variation is a consequence of the granular structure of the film which this gives rise to (unwanted) diffraction contrast in this phase image. This is a significant problem in interpreting magnetic contrast at the smallest scale. A linetrace across the layers shows the magnetic state of the each layer in Fig. 2(b).

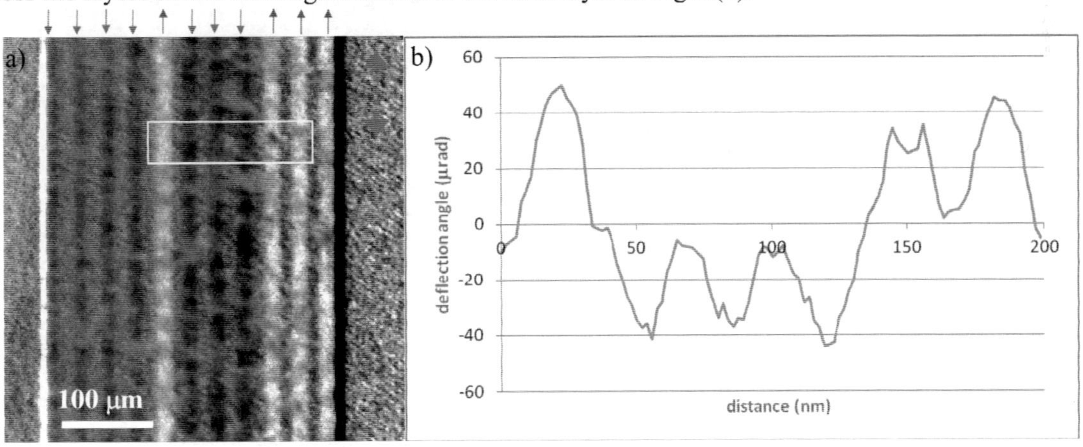

Fig. 2. a) DPC image of exchange biased ferromagnet/antiferromagnet multilayer cross-section. The arrows above the image indicate the direction of magnetisation in the layers and the double headed arrow the direction of sensitivity of the component of magnetic induction. b) Intensity linetrace layers indicated by the yellow box in a) showing the spatial variation of the magnetic induction averaged over this region.

It should be noted we have averaged the signal over a 40 nm width to reduce the effects of diffraction contrast from the granular structure. However even this does not eliminate the effect of diffraction. The linetrace shows a quantitative signal variation between layers corresponding to a deflection of ~ ±40 μrad. This is consistent, but slightly less than expected for a material with an integrated induction of ~ ±80 Tnm, corresponding to permalloy with a saturation induction of 1.0 T and film thickness of 80 nm. The transition between the FM and AF layers shows a variation on a lengthscale of the order of 5 nm. Whilst the average linetrace shows the power of the spatial resolution of the aberration corrected system with the standard DPC detection, the variation resulting from the crystallite structure clearly shows some problems in imaging this type of material. We now turn our attention to the developments in detection whereby pixelated detectors allow a major improvement in magnetic imaging by the DPC technique.

The advantages of using a pixelated detector for Lorentz DPC will be illustrated by examining images obtained from a magnetic domain wall in a polycrystalline magnetic thin film. The material studied here is 20 nm thick thermally evaporated permalloy doped with platinum ($Py_{90}Pt_{10}$) which has a grain size in the range 5-10 nm. By mapping the integrated induction parallel to the wall we obtain an image of a 180° wall using the standard quadrant detector as shown in Fig. 3(a). It is apparent that there is weak, but visible, underlying black/white contrast within the image corresponding to the large scale magnetic domains. However there is also very high contrast on the 5-10 nm scale resulting from differential Bragg scattering in and between the grains.

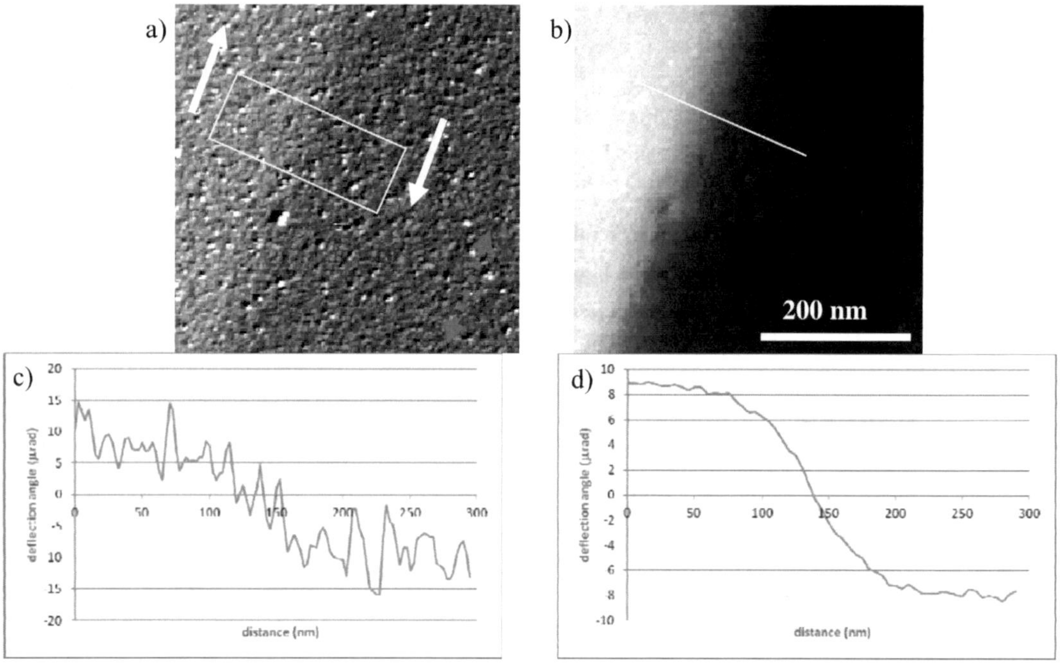

Fig. 3. a) DPC image of 180° domain wall in a 20nm thick platinum doped permalloy film. The arrows indicate the direction of magnetisation in each domain and the double headed arrow the direction of sensitivity of the component of magnetic induction. b) Processed DPC images from a pixelated detector from the same 180° domain wall. c) Intensity linetrace across DPC image of a) averaged over 100 nm of the domain wall as indicated by the yellow box in a). d) Intensity linetrace across DPC image of b) from single line of the domain wall as indicated by the yellow line in a).

When we use the Orius CCD camera to record the bright field disk at each point in the scan we are then able to use image processing to maximize the magnetic phase information that can be extracted. The theory of phase imaging and the contrast functions for DPC imaging are well developed and the theoretical basis is calculated for sinusoidal phase variations being propagated through the electron optical system. The information can be thought of in terms of overlapping disks in the detector plane

[5,6]. In this respect the detector geometry is fundamentally important in terms of any contrast transfer function. The pixelated detector allows considerable flexibility in applying different effective detector geometries to the data sets of the bright field disks. For magnetic imaging using an annular detector which is just smaller than the bright field disk is shown to be very efficient, this is known as modified DPC [5]. This efficiency can be explained simply in that the intensity variations within the bright field disk are mostly associated with the high spatial frequency signal due to the crystallites, by removing this intensity variation this signal is suppressed whilst the signal associated with the beam deflection is preserved. We have found that the most efficient method of removing this variation for the diffraction data set is to use edge detection methods for the bright field disk, which in effect is like having an annular detector just smaller than the disk at each deflection. Put simply this is equivalent to a super efficient modified DPC detector where the ratio k_i/k_α is virtually unity. The improvement in the detected magnetic signal variation is shown Fig. 3(b), where the crystallite signal variation has been massively reduced. Further evidence of the magnitude of the reduction in the signal variation is shown by taking linetraces across the domain wall. For the DPC image the quantitative signal variation based on conversion of the difference signal to a magnetic deflection is shown in Fig. 3(c) with the linetrace for is shown which is averaged over 50 lines. Even this averaging shows a variation associated with the crystallites. By contrast a single line trace across the edge detection processed image is shown on Fig. 3(d). The improvement is quite startling and represents a huge advance in DPC imaging allowing point by point interpretation of magnetic signal variations in polycrystalline films. Whilst the improvement is significant one point of caution must be noted, a quantitative analysis of phase changes where the phase gradient changes rapidly, i.e. at the domain walls needs to be investigated further in order for any quantification to be justified. We will report on current work using the Medipix3 detector in the near future.

5. Conclusions

Aberration corrected DPC STEM offers exciting possibilities for imaging of magnetic materials. The developments in pixelated detection also mean flexible application of the contrast transfer function of the system becomes possible to improving magnetic and more generally phase imaging.

Acknowledgements

We are grateful to colleagues at JEOL, JEOL (UK), Gatan, Deben UK Ltd. and Andrew Armit Designs for their input to the development of the instrument and associated hardware. Additionally we would like to thank Prof R. Bowman and Dr S. O'Reilly for provision of the exchange biased sample, Dr A Hrabec for provision of the PyPt sample and to Mr F. Gonclaves, and Mr W. Smith for cross-section sample preparation. We also acknowledge financial support from the EPSRC (EP/I013520/1), the Scottish Universities Physics Alliance (SUPA), the University of Glasgow and Seagate. The link for open access research data is found here (http://dx.doi.org/10.5525/gla.researchdata.204)

References

[1] Chapman J N 1984 *J. Phys. D. Appl. Phys.* **17** 623
[2] Krivanek O L, Corbin G J, Dellby N, Elston B F, Keyse R J, Murfitt M F, Own C S, Szilagyi Z S and Woodruff J W 2008 *Ultramicroscopy* **108** 179
[3] McVitie S, McGrouther D, McFadzean S, MacLaren D A, O'Shea K J and Benitez M J 2015 *Ultramicroscopy* **152** 57
[4] Ballabriga R, Campbell M, Heijne E, Llopart E, Tlustos E and Wong W 2011 *Nucl. Intsr. Meth. Phys. Res. A* **633** S15
[5] Chapman J N, McFadyen I R and McVitie S 1990 *IEEE Trans. Magn.* **26** 1506
[6] Pennycook T J, Lupini A R, Yang H, Murfitt M F, Jones L and Nellist P D 2015 *Ultramicroscopy* **151** 160
[7] Hill C B, Hendren W R, Bowman R M, McGeehin P K, Gubbins M A and Venugopal V A 2013 *Meas. Sci. Technol.* **24** 045601

Visualisation of high temperature magnetisation states in magnetite grains using off-axis electron holography

Trevor P. Almeida[1]*, Adrian R. Muxworthy[1], András Kovács[2], Wyn Williams[3] and Rafal E. Dunin-Borkowski[2]

[1]Department of Earth Science and Engineering, South Kensington Campus, Imperial College London, London, SW7 2AZ, UK.
[2]Ernst Ruska-Centre for Microscopy and Spectroscopy with Electrons (ER-C) and Peter Grünberg Institute, Forschungszentrum Jülich, D-52425 Jülich, Germany.
[3]School of GeoSciences, University of Edinburgh, Edinburgh, EH9 3JW, UK.

Abstract. The production of a synthetic basalt comprising Fe_3O_4 grains (~ 50 nm to ~ 500 nm), via a glass ceramic method, has been confirmed using transmission electron microscopy and X-ray diffractometry. Off-axis electron holography combined with *in situ* heating allowed for the visualisation of non-uniform vortex states present in saturation remanent structures, and their variation approaching the Curie temperature; determined separately by bulk thermomagnetic measurements.

1. Introduction

The ability of a rock to reliably record the geomagnetic field depends on the geometry and size of the constituent magnetic particles. In the field of paleomagnetism, the stability of the remanent signal is a critical factor for the recovery of reliable intensity and directional information. Small, magnetically uniform grains, termed single domain (SD), are considered the strongest remanence carriers with good stability over geological time periods [1]. However, the magnetic signature in igneous rocks is usually dominated by larger magnetic grains (~ 0.1 – 10 μm) that display non-uniform magnetic structures, commonly termed pseudo-single domain (PSD), as their bulk magnetic characteristics are similar to SD particles. The most common magnetic minerals on Earth fall within the iron-titanium-oxide ($Fe_{3-x}Ti_xO_4$) system [2], and are commonly found in igneous rocks, including mid-ocean ridge basalts that make up much of ocean plates. In particular, magnetite (Fe_3O_4) is a very important magnetic mineral due to its high abundance and dominating magnetization. In igneous rocks, the leading source of magnetic memory by the constituent magnetic minerals is termed thermal remanent magnetization (TRM) and is acquired in the direction of the ambient geomagnetic field as the grains cool below their Curie temperature (T_C (~ 580 °C for magnetite)). However, current understanding of the thermomagnetic behaviour of PSD grains during TRM acquisition is limited and not sufficient to elucidate the intricate details of the stability of PSD states with temperature.

The transmission electron microscopy (TEM) technique of off-axis electron holography permits the nanometre-scale imaging of magnetic induction within and around materials as a function of applied field and temperature [3]. In this context, the synthesis of synthetic basalts containing Fe_3O_4 grains is presented, along with their bulk magnetic properties, and off-axis electron holography is combined with the heating of PSD magnetite (Fe_3O_4) grains *in situ* within the TEM.

*Author to whom correspondence should be addressed: t.almeida@imperial.ac.uk.

2. Experimental

Samples were prepared by a glass-ceramic method [4] using laboratory grade powders of Fe_2O_3, SiO_2, $CaCO_3$, K_2CO_3, and Na_2CO_3, which were mixed with aqueous polyethylene oxide solution and pressed into pellets (~ 5 mm in diameter) on the end of Pt wires. The pellets were suspended in a vertical temperature controlled tube furnace under an oxygen partial pressure and melted at 1400 °C for 15 h to equilibrate Fe^{2+}/Fe^{3+} ratios with the oxygen fugacity, before subsequently quenched in the bottom of the tube furnace. After quenching, the pellets were reheated at 750 °C for 3 h and quenched again. For the purpose of structural characterization, quenched pellets of the synthetic basalt were crushed into a fine powder using a pestle and mortar, and were deposited onto single crystal silicon substrates for crystallographic identification using X-ray diffractometry (PANalytical X'Pert PRO Diffractometer). Conventional bright-field (BF) imaging was performed using a FEI Tecnai TEM operated at 200 kV (ER-C). For the purpose of magnetic measurements, the sample was examined at the Natural Magnetism Group laboratory at Imperial College London. The first-order reversal curve (FORC) and high temperature thermomagnetic curve were measured using a Princeton Measurements Vibrating Sample Magnetometer fitted with a furnace; heating was performed in flowing He.

For the purpose of *in situ* heating TEM investigations, desirable regions were deposited and welded with platinum to contacts on a double-tilt EMheaterchip™ before being ion-milled into thin sections. Off-axis electron holograms were acquired at 400°C, 500 °C and 550°C using an FEI Titan 80–300 TEM in Lorentz mode (ER-C), with a charge-coupled device camera and electron bi-prism operated typically at 90 V. The direction of magnetisation in the particle was reversed at each temperature interval *in situ* in the TEM by tilting the sample ± 75° and turning on the conventional microscope objective lens to apply a magnetic field of ~1.5 T to the sample, parallel to the direction of the electron beam. The objective lens was then turned off and the sample tilted back to 0° for hologram acquisition in field-free conditions with the particles at remanence of saturation magnetisation.

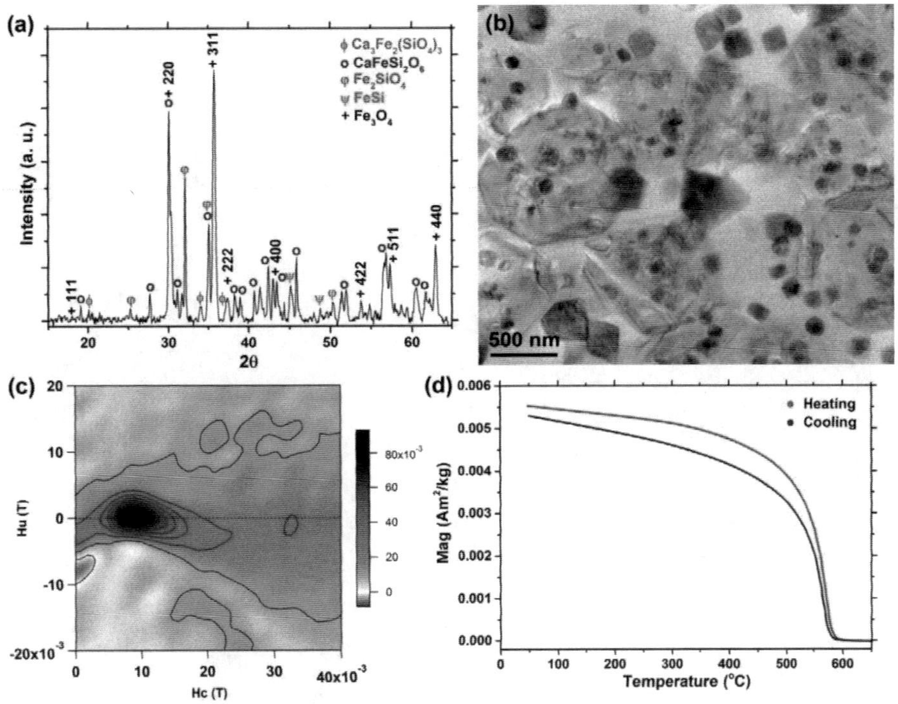

Figure 1 (a) XRD pattern of the synthetic basalt, confirming the Fe_3O_4 phase (indexed, JCPDS 75-449), as well as the additional phases of $CaFeSi_2O_6$, $Ca_3Fe_2(SiO_4)_6$ and FeSi. (b) BF TEM diffraction contrast image of the Fe_3O_4 grains within the silicate matrix ranging from ~ 50 to ~ 500 in diameter. (c) FORC diagram acquired at room temperature (smoothing factor = 7). The measurement time was 250 ms. (d) Thermomagnetic curves of the synthetic basalt. Heating was performed in flowing helium, in a field of 300 mT.

3. Results

The furnace annealing of the precursor powder mixture led to the production of synthetic basalts comprising Fe_3O_4 grains of varied size, the evidence for which is now presented in detail.

Fig. 1 provides information on the size, structure and magnetic properties of the synthesised sample. The peaks in the XRD pattern of Fig. 1a are in good agreement with the presence of Fe_3O_4 (Joint Committee on Powder Diffraction Standards (JCPDS) reference 75-449), as well as pyroxene hedenbergite ($CaFeSi_2O_6$, JCPDS 70-1876), andradite ($Ca_3Fe_2(SiO4)_3$, JCPDS 84-1938), FeSi (JCPDS 86-0795) and Fayalite (Fe_2SiO_4, JCPDS 71-1673). The BF TEM image of Fig. 1b displays Fe_3O_4 grains ranging in size, from ~ 50 nm to ~ 500 nm, along with the surrounding glass ceramic matrix. It is evident that the smaller Fe_3O_4 grains are generally confined within glassy regions, separate from the larger Fe_3O_4 grains. The FORC diagram shows the sample to exhibit behaviour characteristic of a mixture of SD and PSD grains. The negative region close to the vertical H_u-axis indicates that the controlling anisotropy in SD grains is uniaxial, whilst the curvature along the H_c-axis reveals the presence of PSD grains. High-temperature thermomagnetic analysis (Fig. 1d) indicated that the Fe_3O_4 NPs were thermally stable. The T_C was determined from the heating curve using the second-derivative method [5] and calculated as 585 ± 5°C, which is in good agreement with stoichiometric Fe_3O_4.

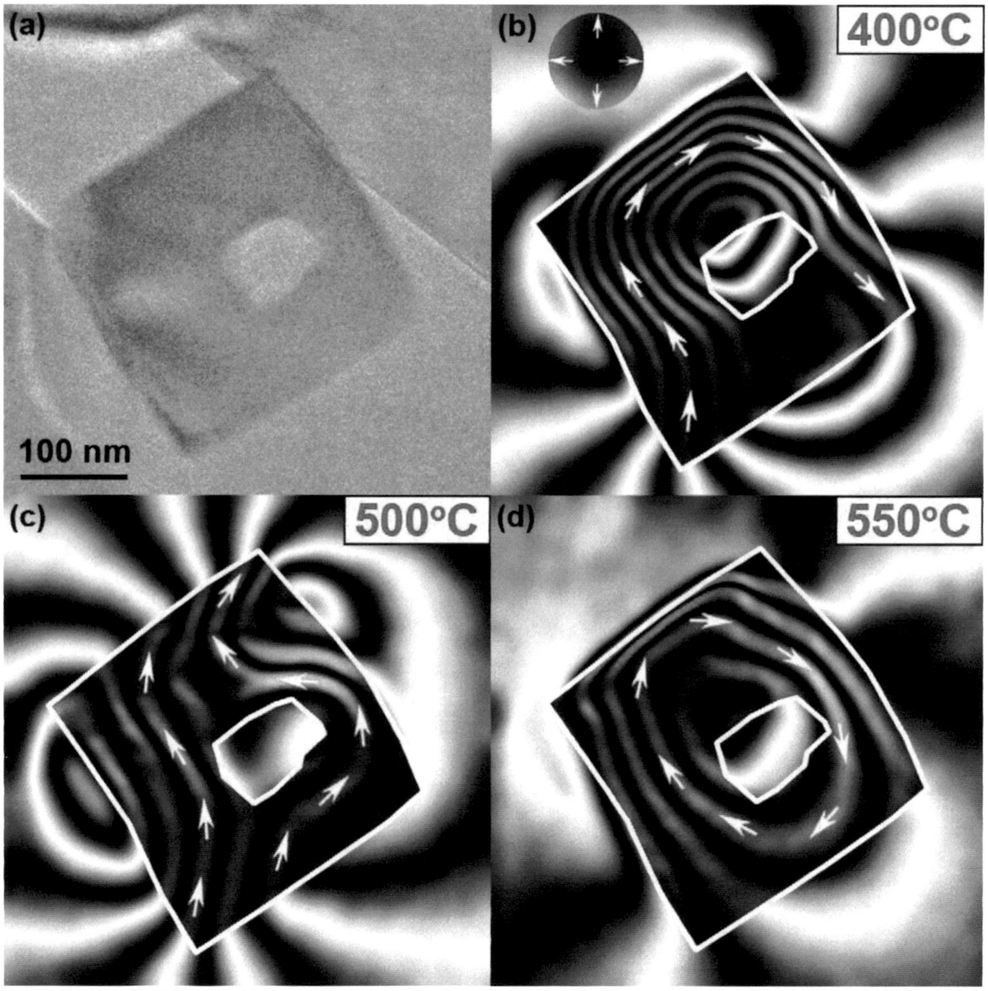

Figure 2 (a) Bright-field TEM image of a square Fe_3O_4 grain within the silicate matrix (~ 300 nm along each side), exhibiting a pore at its centre. (b-d) Magnetic induction maps reconstructed from holograms taken during *in situ* heating to (b) 400°C; (c) 500°C; and (d) 550°C. The contour spacing is 0.39 radians for all magnetic induction maps and magnetisation direction is shown using arrows, as depicted in the colour wheel.

Fig. 2 presents the thermomagnetic behaviour of a Fe_3O_4 grain embedded in a silicate matrix as it is heated to temperatures of 400 °C, 500 °C and 550 °C. The bright-field TEM image of Fig. 2a displays the square Fe_3O_4 grain with sides ~ 300 nm in length and a pore located in its centre. Fig. 2b displays a magnetic induction map of the Fe_3O_4 grain at 400 °C, and reveals its magnetization to resemble a horseshoe-like shape, with the magnetic contours flowing in a clockwise direction, along with a component of stray magnetic field, indicative of a PSD state. An increase in temperature to 500 °C (Fig. 2c) results in a marked change in the magnetisation, with the contours flowing vertically through the Fe_3O_4 grain from its bottom to top, as well as exhibiting a prominent stray magnetic field. In particular, the magnetic contours are observed to curve around the right of the centre pore, before re-joining the contours in the left of the grain. At 550 °C, the magnetization is observed to close into a vortex structure around the centre pore (Fig. 2d), along with a reduced stray magnetic field.

4. Discussion

This combined TEM, XRD and magnetic investigation of the production of synthetic basalts has provided evidence for the growth of Fe_3O_4 grains, ranging from ~ 50 to ~ 500 in diameters, within the silicate matrix. The smaller 50 nm grains are observed to be confined within glassy regions, which have been previously identified as Ca-rich, and these SD domain grains are considered to form during the rapid quenching stage of synthesis [4]. The larger grains, < 500 nm in diameter, are dispersed in-between the Ca-rich regions and suggested to contribute the PSD signal in the FORC diagram. The thermomagnetic curve displays a T_C characteristic of stoichiometric Fe_3O_4, and shows a significant drop in magnetisation above 400 °C, identifying an ideal temperature range for investigating the high temperature behaviour of PSD Fe_3O_4 grains using electron holography.

The magnetic induction maps of Fig. 2b – 2d provide a visual representation of the variation in remanent magnetisation states within the Fe_3O_4 grain as a function of temperature, all induced under a saturating field of ~ 1.5 T applied in the same direction. It is clear that the temperature at which the electron holograms are acquired plays a critical role into which magnetic configuration is energetically favoured. Considering all three magnetisation states are markedly different, the thermal energy imparted on the Fe_3O_4 grain during magnetic signal recording is considered to actively influence the direction of the remanent signal. In the field of palaeomagnetism, this study provides much insight into the role of the effect of temperature on signal acquisition of the ambient geomagnetic field as the grains cool below their T_C, i.e., it indicates that domain states are likely to be highly temperature dependent. In the case of metamorphic rocks, which experience multiple heatings, a grain may be re-magnetised into different states on each heating cycle. Hence, the overall palaeodirectional information from all the grains in a rock will probably be maintained, but due to some grains re-aligning their magnetisation, the signal will decrease in net palaeointensity.

In summary, complementary TEM and XRD have confirmed the synthesis of Fe_3O_4 grains (~ 50 to ~ 500 nm in diameters) within a synthetic basalt. Off-axis electron holography combined with *in situ* heating showed the variation of induced magnetic domain states at elevated temperatures.

References

[1] Néel L 1995 *Adv. Phys.* **4**, *191-242 L.*
[2] Dunlop D J and Ozdemir O, 1997 Rock Magnetism: Fundamentals and Frontiers, (Cambridge Univ. Press, Cambridge, UK).
[3] Dunin-Borkowski R E, McCartney M R, Frankel R B, Bazylinski D A, Posfai M and Buseck P R 1998 *Science* **282,** 1868-1870.
[4] Almeida T P, Muxworthy A R, Williams W, Kasama T and Dunin-Borkowski R E 2014 *Geochem. Geophys. Geosyst.* **15**, 161-175.
[5] Tauxe L 1998 Paleomagnetic Principles and Practice, (Kluwer Acad., Dordrecht, Netherlands).

Acknowledgements

The authors would like to thank the Natural Environment Research Council for funding NERC grant NE/J020966/1. Thanks also go to the ER-C for the use of their microscopy facilities.

The Effect of Cu and Ge Additions on Strength and Precipitation in a lean 6xxx Aluminium Alloy

E A Mørtsell[1, 4], C D Marioara[2], S J Andersen[2], J Røyset[3], O Reiso[3] and R Holmestad[1]

[1]Department of Physics, NTNU, 7491 Trondheim, Norway
[2]SINTEF Materials and Chemistry, N-7465 Trondheim, Norway
[3]Hydro Aluminium R&D Sunndal, N-6600 Sunndalsøra, Norway

E-mail address: eva.mortsell@ntnu.no

Abstract. It has been demonstrated that the strength loss in a lean Al-Mg-Si alloy due to solute reduction could be compensated by back-adding a lower at % of Ge and Cu. Nanosized precipitate needles which are the main cause of strength in these alloys, and material hardness has been correlated to parameters quantified by TEM. It was found that additions of Ge and Cu strongly affect the precipitation process by increasing precipitate density and reducing precipitate size. Investigations of precipitate atomic structure by HAADF-STEM indicated that they contain mixed areas of known phases and disordered regions. A hexagonal Si/Ge-network was found to be present in all precipitate cross sections.

1. Introduction

Aluminium alloys have a vast number of applications, and the 6xxx series in particular is used as extruded and rolled products ranging from car manufacturing to packaging industry. The strength in these age hardenable alloys is caused by the formation of nanosized precipitates shaped like needles growing along the three Al <100> lattice directions [1-9]. The precipitation sequence in ternary Al-Mg-Si alloys is thoroughly characterized and may be given as [1]:

Supersaturated solid solution → atomic clusters → Guinier Preston-zones (pre-β") [2-3] → β" [4-5] → β' [6], *U1* [7], *U2* [8], *B'* [9] → β, Si (stable)

The precipitate needles consist of combinations of the added alloying elements and Al itself, and different precipitate types are found for different alloys and after varying heat treatments. It is essential to find the combinations of solute additions and heat treatment resulting in an optimal balance between strength and production efficiency. Pure aluminium would give the fastest possible extrusion of aluminium alloys [10]. However, the strength of pure aluminium is too low for most applications. In this work, the total amount of Mg and Si in a 6060 Al alloy has been reduced and a lower atomic percentage of Ge and Cu has been back-added. The primary goal of the project has been to maintain initial strength, while increasing extrusion speed because of the lower amount of solutes.

Aberration-corrected High Angle Annular Dark Field Scanning Transmission Electron Microscopy (HAADF-STEM) was chosen as the main characterization technique of the precipitate

[4] To whom any correspondence should be addressed.

atomic structures. Compared to the conventional High Resolution TEM (HRTEM) technique, aberration-corrected HAADF-STEM has several advantages. In addition to a high spatial resolution approaching 0.1 nm, the Z-contrast images are more directly interpretable, being less affected by small changes in objective lens defocus and specimen thickness [11, 12]. With this technique the heavy Ge and Cu atomic columns can be easily identified in the precipitate cross sections.

2. Methods/Materials

An Al 6060 alloy was used as a dense reference, from here on denoted RX0. The second alloy, RXGC1, contained Cu and Ge in addition to the common 6xxx alloying elements. A lean reference, RX1, with reduced Mg and Si contents, directly comparable to RXGC1 in at %, was used as a lean reference alloy. The chemical compositions of the alloys are given in Table 1.

Table 1. Nominal alloy composition in at %. Total effective solute is given in the "total solute"-column.

Alloy	Si	*Ge*	Mg	*Cu*	Fe	*Mn*	*Total Solute*
RX0	0.43	-	0.41	-	0.10	0.015	0.80
RXGC1	0.36	0.02	0.36	0.021	0.10	0.015	0.70
RX1	0.36	-	0.36	-	0.10	0.015	0.66

The alloys were cast, homogenized, extruded and heat treated identical to the description in [13].

TEM investigations have been done for samples artificially aged for 4 hours at 195 °C. Foils for TEM specimens were made by polishing slices from the sample blocks down to a thickness of approximately 100 μm. The foils were electropolished with an electrolyte consisting of one third nitric acid and two thirds methanol kept at -25 °C by adding liquid nitrogen to the solution. The HAADF-STEM images presented in this work were recorded with a double C_s corrected JEOL ARM200F, operated at 200 kV. The probe size was 0.08 nm and the inner collection angle was 50 mrad. For precipitate statistics computed using a methodology described in [14], a Philips CM30 TEM operated at 150 kV and equipped with a Gatan PEELS system was used.

3. Results/Discussions

Statistical results of precipitate microstructure are given in Table 2. It is clear from Table 2 that RXGC1 almost compensates for the strength loss in HV upon removing some Mg and Si. RXGC1 has the same amount of solute as RX1, but the hardness is remarkably higher. This increase in HV can partly be explained by the refinement of microstructure in RXGC1, as compared to RX0. The needles have significantly smaller cross sections and lengths in RXGC1, and in addition the number density has increased by approximately a factor of 50 when comparing to the lean reference RX1. It becomes clear that a lower volume fraction of needles is needed in RXGC1 to achieve approximately the same hardness as for the dense reference RX0.

RX0 does not have any noteworthy amount of disordered precipitates, and β" is the main hardening precipitate. An example of a typical β" cross section in RX0 is shown in Fig. 1 (a), in agreement with [4].

HAADF-STEM revealed RXGC1 to consist of partly disordered precipitates, where the projected hexagonal Si/Ge - network can be observed in all cross sections [10]. Z-contrast confirms a clear trend of Ge replacing or substituting Si network columns, see Fig. 1 (b) and Fig. 2. The strong atomic number dependence of HAADF STEM makes it possible to distinguish Ge (Z = 32) columns and Mg, Al, Si or Cu columns (Z = 12, 13, 14 and 29 respectively). Intensity peaks in the HAADF STEM images correspond directly to atomic column positions.

Table 2. Estimated average precipitate cross sections (CS), needle lengths, number densities, volume fractions and measured Vickers hardness (HV) for each condition. The volume fractions of precipitates are given as 95 % confidence intervals [13]

Alloy	CS [nm^2]	Needle Length [nm]	Number Density [#/μm^{-3}]	Volume Fraction	HV
RX0	24 ± 2	95 ± 5	3200 ± 400	{0.64, 0.80}	77.0
RXGC1	12 ± 1	57 ± 8	8300 ± 1300	{0.48, 0.62}	74.5
RX1	42 ± 4	243 ± 73	170 ± 50	{0.12, 0.19}	50.0

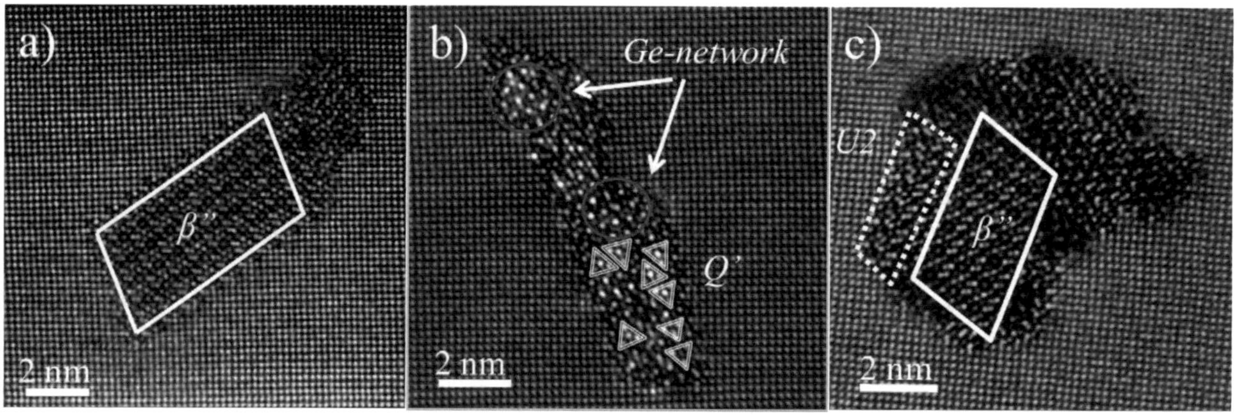

Figure 1. HAADF STEM images of precipitates from the three investigated alloys. The precipitates chosen are typical for the respective alloy and shows that the addition of Ge and Cu strongly affects formation of the phases. The images are taken from a) RX0, b) RXGC1 and c) RX1. In b) the hexagonal Ge-network is indicated with red circles and fragments the Q' phase are indicated with white triangles. In a) and c) the β'' phase is marked with white parallelograms. Part of a U2 phase is also indicated in c) with white dashed lines.

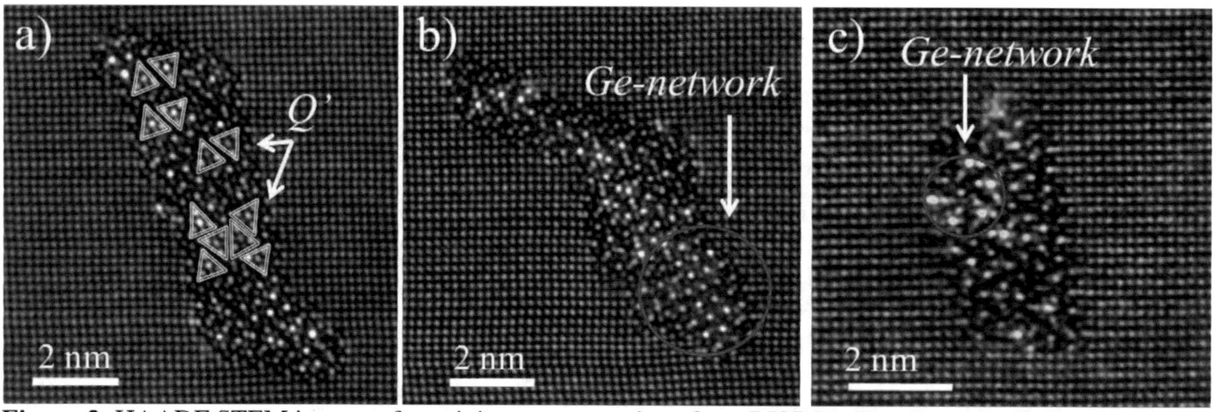

Figure 2. HAADF STEM images of precipitate cross sections from RXGC1. The Ge-network is indicated with arrows and red circles, it is also present in other areas than those pointed out. The Q' fragments are pointed out with arrows and white triangles. The precipitates in (a) and (b) were found along a dislocation line.

Ge and Cu were found to be present in all precipitate cross sections in RXGC1, so it is clear that these elements play an important role in the precipitation process. Since Ge and Cu are often located also in the centre of the precipitate cross sections it is feasible that they should be active in the nucleation process. As in denser Ge – and Cu containing Al-alloys [15], the Q' phase form frequently also in this leaner alloy (Fig. 2), however, not mixed with the Ge-network.

The cross sections observed in the lean reference alloy RX1 are few and large, see Table 2. An example is given in Fig. 1 (c). The cross sections in this alloy can typically be identified as combinations of several phases. It is usual to find phases associated with overaging of the material here, such as U2 and β'. It should be noted that scan distortions and drift are present in some of the HAADF STEM images. The distortions can in particular be observed in Fig. 2 (b), where the Al lattice has a wave-like appearance in some areas.

4. Conclusions

It has been shown that small additions of Ge in combination with Cu have a strong influence on precipitation in lean 6xxx alloys, leading to high densities of fine precipitates and high hardness values. A lower volume fraction of solute was needed for the Ge- and Cu added alloy in order to obtain the same hardness values as the reference.

The precipitates in the Ge-Cu added alloy were mainly disordered with a hexagonal ordering of Ge/Si columns. Most precipitates also incorporated fragments of the Cu-containing Q' phase. It is clear that Ge and Cu play an important role in the precipitation process.

References

[1] Marioara C D, Nordmark H, Andersen S J, Holmestad R 2006 *J. Mater. Sci.* **41**, pp 471 - 478
[2] Edwards G A, Stiller K, Dunlop G L, Couper M J 1998 *Acta Mater.* **46**, pp 3893 – 3904.
[3] Marioara C D, Andersen S J, Jansen J, Zandbergen H W 2001 *Acta Mater.* **49**, 321-328.
[4] Andersen S J, Zandbergen H W, Jansen J, Træholt C, Tundal U, Reiso O 1998 *Acta Mater.* **46**, pp 3283 – 3298.
[5] Hasting H S, Frøseth A G, Andersen S J, Vissers R, Walmsley J C, Marioara C D, Danoix F, Lefebvre W, Holmestad R 2009 *J. Appl. Phys.* **106**, 123527
[6] Vissers R, van Huis M A, Jansen J, Zandbergen H W, Marioara C D, Andersen S J 2007 *Acta Mater.* **55**, pp 3815 – 3823.
[7] Andersen S J, Marioara C D, Frøseth A, Vissers R, Zandbergen H W 2005 *Mater Sci. Eng.* A **390**, pp 127 – 138.
[8] Vissers R, Marioara C D, Andersen S J, Holmestad R 2008 *Aluminium Alloys* **2**, pp 1263 – 1269.
[9] McLeod D 1962 *Mechanical Properties of Metals* (John Wiley & sons Inc, New York) p 162.
[10] Andersen S J, Marioara C D, Vissers R, Zandbergen H W 2003 *Inst. Phys. Conf. Ser.* **179**, pp 225 – 228.
[11] Nellist P D, Pennycook S J 2000 *Elsevier* **113**, pp. 147 – 203.
[12] Yamazaki T, Kawasaki M, Watanabe K, Hashimoto I, Shiojiri M 2002 *Ultramicroscopy* **92** pp. 181 – 189.
[13] Mørtsell E A, Marioara C D, Andersen S J, Reiso O, Røyset J, Holmestad R 2015 *Met. Trans.* A
[14] Marioara C D, Andersen S J, Zandbergen H W, Holmestad R 2005 *Metall. Mat. Trans.* A, **36**, pp. 691 – 702.
[15] Bjørge R, Marioara C D, Etheridge J 2013 *Imaging and Microscopy*, **4**, pp 39.

Sub-surface characterisation of tribological contact zone of metal hip prostheses

P Zeng[1], J Sharp[1], A Rana[1], R Thompson[1], W M Rainforth[1] and R B Cook[2]

[1] Department of Materials Science and Engineering, University of Sheffield, UK
[2] National Centre for Advanced Tribology at Southampton, University of Southampton, UK

E-mail: p.zeng@sheffield.ac.uk

Abstract. Many metal on metal joint failures have been associated with adverse local tissue reactions due to the response of the body to wear debris and corrosion products, released from the bearing surface or the taper interface. Both a carbonaceous layer and an oxide film are reported on the CoCrMo metal joints, which play important roles in the wear resistance of the material. However, there is a lack of quantitative data on the structure and distribution of the carbonaceous layer and the oxide film. The current work provides a detailed investigation of the surface damage on retrieved CoCrMo taper interfaces. In addition, systematic differences in the starting surface structure of biomedical CoCrMo were studied through a comparison of a standard mechanical polished (MP) and an electropolished (EP) surface after reciprocating test. Site-specific FIB/TEM cross-sections show the evidence of the carbonaceous layer and the oxide film on both CoCrMo taper interfaces and MP, EP CoCrMo, ranging from 5 -500 nm depending on the location. The amorphous carbonaceous layer exhibits a π * peak in the EEL spectra, with trace of Ca and N. The oxide film on the taper interfaces has a porous texture and HRTEM indicates chromium oxide nano-crystals in an amorphous background, however, only a very thin oxide film (~ 2 nm) exists on MP and EP CoCrMo after reciprocating test.

1. Introduction

Metal debris and metal ions released from CoCrMo alloys is the subject of considerable interest due to the higher than expected failure rate of metal on metal (MoM) hip replacements [1, 2]. The Medicines and Healthcare Products Regulatory Agency (UK) and the Food and Drug Administration (US) issued alerts for all MoM hip replacement in 2010 and 2011, respectively, with concerns on the adverse local tissue reactions (ALTR) to metal ions, resulting dramatic drop of the MoM market in recent years. Metal ions and debris release is not only derived from the bearing surfaces, but can also arise from the taper interface between the head and stem components of modular total hip replacements [3]. Release of material from this interface has been linked to ALTR from MoM and Metal on Polymer (MoP) joints. It is well know that an oxide film is present on the surface of CoCrMo and plays an important role in the corrosion resistance of CoCrMo alloy. In addition, a tribo-layer due to tribochemical reactions has been widely observed in MoM prostheses by several authors [4, 5]. Despite the importance of these layers, there has been a relative lack of quantitative data on the structure of these layers. Furthermore, the effect of the starting structure of the CoCrMo alloy on the subsequent wear behaviour has not be established. In the current study, detailed analyses of near surface layers on the taper interfaces, as well as the biomedical grade CoCrMo with different surface finish (MP and EP) under reciprocating wear test, are provided using extensive site-specific transmission electron

Journal of Physics: Conference Series **644** (2015) 012029 doi:10.1088/1742-6596/644/1/012029

microscopy (TEM) with associated high resolution transmission electron microscopy (HRTEM) and electron energy loss spectroscopy (EELS).

2. Experimental procedure

The taper interfaces of two retrieved femoral heads were studied. The two femoral heads were a 40 mm (Case1) and a 44 mm (Case 2) diameter zero offset low Friction Ion Treatment (LFIT^{TM}) CoCrMo anatomic heads which were both retrieved after 4 years *in vivo*. The bearing was a MoP articulation against a zero degree X3 polyethylene liner in a Trident HA (hydroxyapatite) coated acetabular shell made of Ti6Al4V ELI (Extra Low Interstitial). The stem was an Accolade Femoral stem, made of a β-titanium alloy (TMZF: titanium, molybdenum, zirconium and iron). Ethical approval was granted by the National Research Ethics Service Committee South Central-Southampton A. The samples were sterilised and stored separately in 10% buffered formalin.

The cast biomedical grade CoCrMo alloy of the reciprocating wear test was obtained from Finsbury Orthopaedics Ltd (UK). The chemical analysis revealed that the material matched the specifications required for the ASTM F75 standard. All CoCrMo were subjected to a mechanical polish, including those to be electro-polished. The electro-polishing was undertaken by a specialist electro-polishing company (Anopol Ltd., Birmingham, UK). Mechanical polished and electro-polished CoCrMo, namely MP and EP CoCrMo, were under reciprocating sliding test performing on a reciprocating ball-on-plate UMT tribometer (Bruker, USA), using 25 vol% bovine serum albumin (BSA) solution at room temperature. A wear track length of 10 mm, reciprocating at 600 rpm (02 m/s) under various loads (i.e. 2N and 4N) were chosen as the standard test parameter.

Site-specific TEM cross-sections of taper interfaces and wear tracks of MP and EP CoCrMo were prepared using a Quanta 200D FIB-SEM (FEI, the Netherlands). HRTEM with Cs corrector (Jeol R005, Jeol, Japan) in conjunction with electron energy loss spectroscopy (EELS, Gatan, USA) was employed to analyse the near surface structure.

3. Results and discussion

The initial surface finish was different on the taper interfaces and the MP, EP surfaces. A rough surface exhibiting the machining marks from the manufacture process could be seen on the taper interfaces. The EP CoCrMo has a highly polished surface free of residual deformation, whereas the MP CoCrMo has a highly polished surface with residual deformation comprising ε-martensite plates, mechanical twins and a high dislocation density. Despite the difference in surface finish, both carbonaceous layer and oxide film were observed from the taper interfaces and the MP, EP CoCrMo, as shown in Fig.1a and Fig.1b (sub-layer in the carbonaceous layer in Fig.1b is due to re-deposition of Pt). The thickness of the carbonaceous layer varied from region to region on the taper interfaces, ranging from 20 nm to 500 nm. A difference in the thickness of the carbonaceous layer was also seen on the MP and EP CoCrMo after reciprocating test at various loads, with ~ 30 nm for MP-2N, ~500 nm for MP-4N, ~ 40 nm for EP-2N and ~ 10 nm for EP-4N. Interestingly a thinner carbonaceous layer was seen for EP-4N compared with EP-2N. In addition, the carbonaceous layer on the taper interfaces and the MP CoCrMo samples was detectable. The carbonaceous layer was predominantly amorphous but with some small crystals in range of ~ 2 nm were observed (as shown in Fig.1c and circles in Fig.1e). Those nano-crystals were believed to be chromium oxide trapped in the carbonaceous layer.

The oxide film observed on the taper interfaces showed a porous structure (Fig.1a), ranging from 10 - 100 nm from region to region. HRTEM image (Fig.1d) showed nano-crystals ~ 2 nm embedded in the amorphous material. These nano-crystals were confirmed to be chromium oxide by EELS (spectrum 4 in Fig.2). A thin layer of oxide film of ~ 2 nm was observed from the MP and EP CoCrMo sample after reciprocating test, as shown in Fig.1e. HRTEM image indicated that the oxide film was a thin layer of nano-crystals and EEL spectrum (spectrum 3 in Fig.1) confirmed they were chromium oxides.

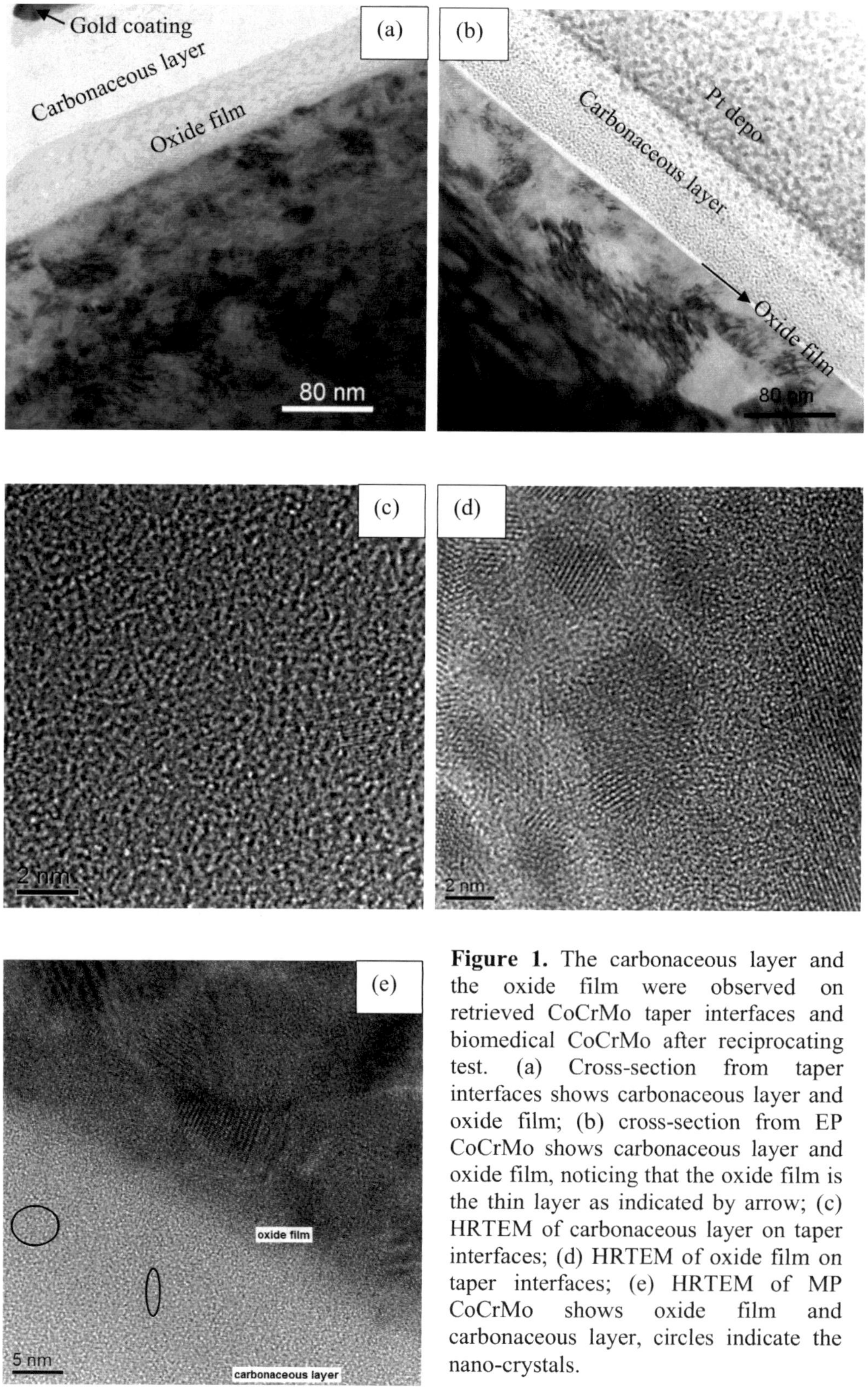

Figure 1. The carbonaceous layer and the oxide film were observed on retrieved CoCrMo taper interfaces and biomedical CoCrMo after reciprocating test. (a) Cross-section from taper interfaces shows carbonaceous layer and oxide film; (b) cross-section from EP CoCrMo shows carbonaceous layer and oxide film, noticing that the oxide film is the thin layer as indicated by arrow; (c) HRTEM of carbonaceous layer on taper interfaces; (d) HRTEM of oxide film on taper interfaces; (e) HRTEM of MP CoCrMo shows oxide film and carbonaceous layer, circles indicate the nano-crystals.

The thickness and the structure difference of the oxide film on the taper interfaces and the MP, EP CoCrMo could be due to the surface finish. On the taper interfaces, the initial oxide film was thick because of the rough surface finish, which acted as the contact zone and the tribochemical reactions could be the reason to form later thick and porous oxide film. However, with a highly polished surface finish, there was a thin/no oxide film on as-polished surface, which led to a thin oxide layer (~ 2 nm) on the MP and EP CoCrMo after reciprocating test.

The EEL spectrum from the carbonaceous layer on the taper interface (spectrum 1 in Fig.2) showed C-K, Ca-L, N-K and O-K edges. No Cr-L edges were observed, even though small (<2nm) chromium oxide particles were present inside the carbonaceous layer (Fig.1c). The carbonaceous layer on the MP CoCrMo gave a similar result, spectrum 2 in Fig.2 although the Ca-L peak was more pronounced in this case. The oxide film on the MP CoCrMo, EEL spectrum (spectrum 3 in Fig.2) showed not only O-K and Cr-L edges, but also C-K, Ca-L and N-K edges. The oxide film was only ~ 2 nm such that the spectrum probably contained a signal from the nearby carbonaceous layer. Spectrum 4 in Fig.2 was from the oxide film on the taper interface, showing C-K, O-K and Cr-L edges. A π * peak was observed in all C-K edges.

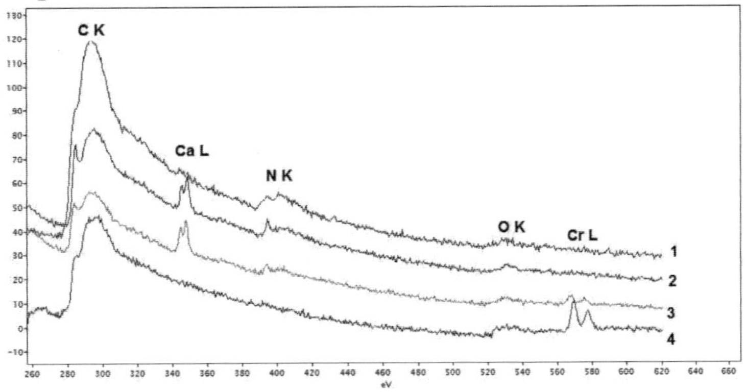

Figure 2. EEL spectra of carbonaceous layer on taper interface (1), oxide film on MP CoCrMo (2), carbonaceous layer on MP CoCrMo (3) and oxide film on taper interface (4).

4. Conclusions

The electropolishing produced a surface free of residual deformation, whereas the mechanical polishing resulted in significant residual deformation in the surface, as seen in the taper interfaces and MP CoCrMo after reciprocating test. Both carbonaceous layer and oxide film were observed from the taper interfaces and the MP and EP CoCrMo after reciprocating test. The amorphous carbonaceous layer exhibit a π * peak in the EEL spectra, with trace of Ca and N. In addition, the carbonaceous layer on the taper interfaces and the MP CoCrMo after reciprocating test were thick and detectable, whereas a thin carbonaceous layer was seen on the EP CoCrMo after reciprocating test. The oxide film on taper interface has a porous structure and HRTEM indicated chromium oxide nano-crystals in an amorphous background, however, only a very thin oxide film (~2 nm) exists on MP and EP CoCrMo after reciprocating test.

References

[1] Smith A, Deippe P, Vernon K, Porter M and Blom A 2012 *Lancet* **379** 1199-1204
[2] De Steiger R N, Hang J R, Miller L N, Graves S E, Davidson D C 2011 *J. Bone. Joint. Surg. Am.* **93** 2287-93
[3] Garbuz D S, Tanzer M, Greidanus N V, Masri B A, Duncan C P 2010 *Clin. Orthop. Relat. Res.* **468** 318-25.
[4] Zeng P, Rainforth W M, Cook R B 2015 *Tribol Int* **89** 86-96.
[5] Zeng P, Rana A, Thompson R, Rainforth W M 2015 *Wear* **332-333** 650-61.

Transmission electron microscopy of a model crystalline organic, theophylline

J Cattle[1], M S'ari[1], N Hondow[1], P Abellán[2], A P Brown[1], R M D Brydson[1]

[1]Institute for Materials Research, School of Chemical and Process Engineering, University of Leeds, Leeds, LS2 9JT, United Kingdom
[2]SuperSTEM, STFC Daresbury Laboratories, Keckwick Lane, Warrington, WA4 4AD, United Kingdom

E-mail: sm09jec@leeds.ac.uk

Abstract. We report on the use of transmission electron microscopy (TEM) to analyse the diffraction patterns of the model crystalline organic theophylline to investigate beam damage in relation to changing accelerating voltage, sample temperature and TEM grid support films. We find that samples deposited on graphene film grids have the longest lifetimes when also held at -190 °C and imaged at 200 kV accelerating voltage. Finally, atomic lattice images are obtained in bright field STEM by working close to the estimated critical electron dose for theophylline.

1. Introduction

It is well known that organic and biological materials are particularly sensitive to electron beam exposure [1]. This makes transmission electron microscopy (TEM) of such materials a challenge, but not entirely impractical. If damage can be mitigated, TEM analysis may be used to acquire useful structural, compositional and even chemical information [2, 3]. Additionally, TEM analysis of certain samples can be used to fill gaps in knowledge left by techniques more typical for organic materials analysis (X-ray diffraction (XRD), polarised light microscopy, atomic force microscopy etc.) For example, conventional TEM (CTEM) was recently used to identify a previously unidentified form of theophylline [4], which only exists as a minor phase (<1%) of form II of theophylline. The ability of TEM to be used for the analysis of individual crystals of material with wide ranges in size is one of its biggest advantages for pharmaceutical product analysis.

There are numerous strategies that can be used to limit the effect of the electron beam on the sample. Generally, the simpler option is to control spot size and other TEM parameters to reduce the intensity of the beam, though there will be a point where the signal will be so low that navigation and imaging of the sample becomes impractical. Here, we present a set of experiments on theophylline aimed at improving its resistance to beam damage without altering the electron beam fluence rate. Theophylline was chosen as a model material for this study. It is a xanthine derivative with the chemical formula $C_7H_8N_4O_2$. It bears many similarities to caffeine and is used as a treatment for conditions such as asthma, chronic obstructive pulmonary disease and sleep apnea [5]. There are many polymorphs of theophylline, with form II being used in this study as it is easily reproduced and readily identifiable by crystal morphology and electron diffraction [4, 6]. Here we perform a thorough test of parameters including accelerating voltage, sample temperature and support film type on theophylline in order to identify the effects of each in electron beam-induced damage.

2. Experimental

A theophylline solution was prepared by dissolving powder form theophylline (purchased from Sigma-Aldrich) in nitromethane. Samples for use in TEM were prepared by pipetting a droplet of the theophylline saturated nitromethane solution on to carbon coated copper TEM grids followed by heating to ~50 °C to evaporate the nitromethane. Large, thin, triangular crystals of form II theophylline (typified in Fig. 1 (a)) were deposited on the grids, with the thinnest axis laying flush with the carbon film, leaving them in an ideal orientation for electron beam analysis.

Analysis was carried out using a FEI Tecnai F20 field emission TEM operated between 80 kV and 200 kV as required, with a tip extraction voltage of 4.5 kV and with an attached Gatan Orius SC600A CCD camera. Gatan DigitalMicrograph software was used for image capture and analysis. Electron fluence rates were calculated using conversion formulae provided by FEI and were dependent on magnification, beam spread, camera emulsion setting and fluorescent plate exposure time. Magnification (2,500 times) and beam spread (0.08 m diameter at the viewing screen) were kept constant, with changes to spot size being used to alter the electron fluence rate. This was kept in a typical range of ~ 10 to ~ 20 $e^{-}nm^{-2}s^{-1}$. Beam damage was analysed by locating a fresh area of crystal in bright field mode followed by capture of diffraction patterns at set time intervals (typically 30 s) until all diffraction spots in the pattern had faded. Post-capture, these series of diffraction patterns were analysed by measuring the intensity of pairs of Bragg spots (typically the {011} spots, indicated in Fig. 1b) at each time interval and relating that to the total electron fluence at that time. Intensities were averaged and normalized against the highest intensity. Critical electron dose values were obtained by measuring the fluence at 1/e of the highest recorded intensity.

Standard conditions used were: 200 kV accelerating voltage at an ambient temperature with the sample on a copper grid with a continuous carbon film (Agar Scientific). Variations on these standard conditions were investigated which included using 80 kV accelerating voltage, cooling the sample to -190 °C and using a holey carbon film and graphene film coated copper TEM grid (Graphene Supermarket).

3. Results and Discussion

3.1. Analysis under standard conditions

Initially, the standard conditions were used to acquire bright field images of theophylline and to produce a benchmark set of diffraction patterns and spot intensity data. A typical image of theophylline form II exhibits the expected triangular plate morphology of the crystals, as well as showing numerous bend contours (Fig. 1a). Such contours indicate crystallinity; they shift over the surface of the crystals as damage or reorientation occur, before disappearing completely upon loss of crystallinity. Fig. 1b shows a typical diffraction pattern ([100] zone axis) for form II theophylline and Fig. 1c shows how the pattern appears after exposure to ~ 4500 $e^{-}nm^{-2}$ total electron fluence. The {011} spots from the [100] pattern are typically used for intensity measurements. The critical dose for theophylline form II under standard conditions is determined to be ~ 2500 $e^{-}nm^{-2}$ (Fig. 2, data set in green). Fluctuation in this and some of the other data sets has been attributed to reorientation or tilt of the crystals as damage progresses, causing spots to brighten or dim.

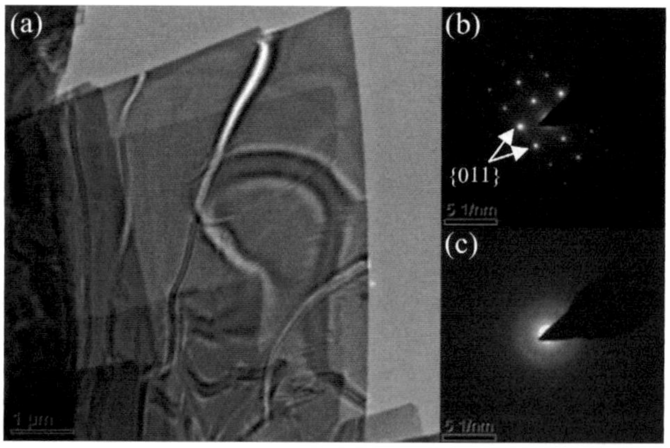

Fig. 1: (a) Bright field TEM image of crystals of theophylline, displaying characteristic triangular, plate-like morphology and multiple bend contours. (b) A typical diffraction pattern along the [100] axis of an unaltered theophylline form II crystal. (c) The same pattern after exposure to ~ 4500 e$^-$nm^{-2}.

3.2. Sample cooling

Analysis of theophylline at -190 °C was carried out using a Gatan cryo-holder, with other variables remaining as per the standard conditions, resulting in a slight increase in critical dose to ~ 2800 e$^-$nm^{-2} (Fig. 2, data set in pink). The lower temperature helps to prevent sample heating, immobilises atoms/molecules and decrease ambient thermal energy that may contribute to bond breaking [1,7].

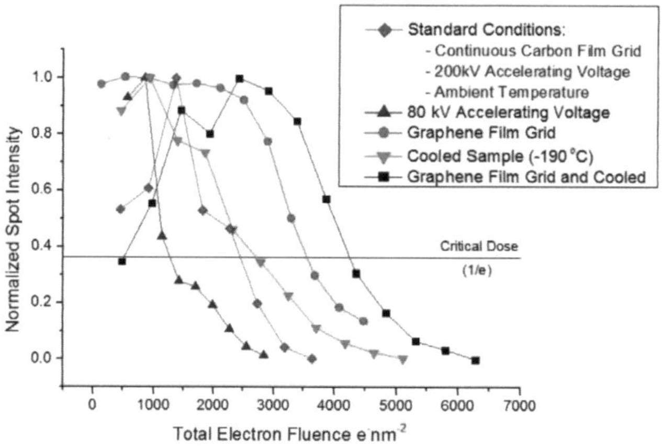

Fig. 2: Graph of normalized {011} spot intensities versus total electron fluence for theophylline samples under a variety of microscope and specimen conditions.

3.3. 80 kV analysis

The result of dropping the accelerating voltage to 80 kV (other variables as per standard conditions) is a significant decrease in critical dose to ~ 1400 e$^-$nm^{-2} (Fig. 2, data set in blue). Reducing the accelerating voltage causes an increase in the beam-sample interaction, most notably the ionisation cross-section, thus decreasing the damage threshold [8].

3.4. Graphene film analysis

Using the standard conditions and theophylline deposited over grids with holey carbon and graphene films resulted in an increase in critical dose to ~ 3600 e$^-$nm^{-2} (Fig. 2, data set in red). Graphene is well known as an electronic and thermal conductor [9], and its effect here is to improve conduction away from the theophylline crystals, preventing heat and charge build up to improve theophylline's stability

Journal of Physics: Conference Series **644** (2015) 012030 doi:10.1088/1742-6596/644/1/012030

under the electron beam. Analysis was also carried out on a sample of theophylline on a graphene film at -190 °C to test if the effects of these would be compounded and resulted in a larger critical dose of ~ 4250 $e^{-}nm^{-2}$ (Fig. 2, data set in black).

3.5. Prospects for direct lattice imaging of theophylline

Scanning TEM (STEM) has previously been shown to be a viable method of obtaining lattice images of sensitive materials [10], as the scanned nature of the focussed probe mitigates beam damage better than the parallel illumination of CTEM. Fig. 3 shows a high resolution bright field phase contrast STEM image of theophylline recorded using a Nion UltraSTEM 100 at 100 kV and a total fluence (~ 9300 $e^{-}nm^{-2}$) three times higher than that usable in CTEM. (100) lattice fringes are barely visible in the raw image, but are more clearly seen in the Fourier filtered image (Fig. 3d).

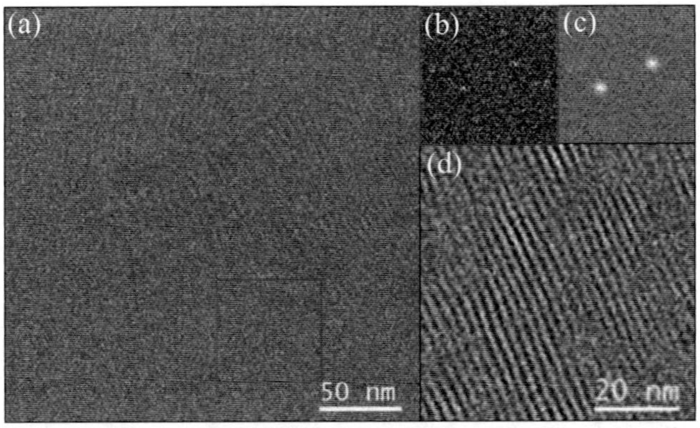

Fig. 3: (a) Bright field phase contrast STEM image of a theophylline crystal. (b) Fourier transform (FFT) of highlighted area in (a). (c) The same FFT with a mask applied to increase the (100) spot intensities by a factor of 10. (d) The inverse of (c), showing the (100) lattice spacings.

4. Conclusions

These results have confirmed the dominance of radiolysis as the main damage mechanism affecting crystals of theophylline when exposed to an electron beam. It has also been shown that this damage can be mitigated by increasing the accelerating voltage where possible, use of a graphene film coated TEM grid and use of a liquid nitrogen cooled sample holder. Bright field STEM shows distinct potential for direct lattice imaging of electron beam sensitive materials.

Acknowledgments

The authors would like to thank Mark Eddleston and Bill Jones (University of Cambridge) for their input and discussion regarding this study.

References

[1] Egerton R F 2013 *Ultramicroscopy* **127** 100
[2] Brydson R et al. 2014 *Journal of Physics: Conference Series* **522** 012060
[3] Fryer J R and Holland F 1984 *Proceedings of the Royal Society of London A* **393** 353
[4] Eddleston M D et al. 2013 *Chemistry A European Journal* **19** 7883
[5] Schultze-Werninghaus G and Meier-Sydow J 1982 *Clinical Allergy* **12** 211
[6] Seton L et al. 2010 *Crystal Growth & Design* **10** 3879
[7] Sader K et al.2009 *Ultramicroscopy* **110** 43
[8] Williams D B and Carter C B 1996 *Transmission Electron Microscopy: A Textbook for Materials Science* vol 2 (New York: Springer)
[9] Pantelic R S et al. 2011 *Nano Letters* **11** 4319
[10] Haruta M and Kurata H 2012 *Scientific Reports* **2** 252

Effects of low-spatial-frequency response of phase plates on TEM imaging

C J Edgcombe
TFMM Group, Department of Physics, Cambridge University, JJ Thomson Ave, Cambridge CB3 0HE, UK

Email@ cje1@cam.ac.uk

Abstract
Images of simple objects produced by a perfect lens and a phase plate have been calculated by use of Abbe theory for Foucault, Hilbert and Zernike phase plates. The results show that with a Zernike plate, white outlines and ringing like those observed previously can be caused by the beam hole, which limits the low-spatial-frequency response of the system even when the lens behaves perfectly. When the change of phase added by the phase plate is distributed over a range of radius rather than a simple step, the unwanted effects are substantially reduced.

1. Introduction

A phase plate in the back focal plane (BFP) of a TEM objective lens can improve contrast at focus for phase objects but may also produce some unwanted effects. Hilbert plates give an impression of adding shadows to an object or differentiating it [1]. Zernike plates may produce white outlines and sometimes 'ringing' of intensity around large objects [2, 3]. Detailed modelling of the response of an objective lens to a single spatial frequency does not show why this behaviour occurs. However, when the effect of a lens and phase plate on an object wave is modelled by including all spatial frequencies and the transverse extent of the object, it becomes clear that the observed defects can be caused by the response of the phase plate to low spatial frequencies, and in particular by its profile closest to the electron beam. The behaviour of interest can be calculated using basic wave-optical theory and with the simple assumption of perfect lens behaviour.

2. Analysis for plate and perfect lens

The image of a physical object depends on more than the contrast transfer function of a single spatial frequency. Such an object contains a spectrum of frequencies and also extends over a transverse area. The image can be found by using Abbe's theory of imaging [4]:

- The object's phase distribution $f(\boldsymbol{x}_o)$ has a spectrum $F(\boldsymbol{q})$ in angular spatial frequency $\boldsymbol{q}$;
- The lens collects rays of the same $\boldsymbol{q}$ at a single transverse coordinate $\boldsymbol{x}_p$ (or $\boldsymbol{r}_p$) in the BFP;
- The phase plate multiplies the spectrum by a transmission function T which is a function of x_p (or $\boldsymbol{r}_p$) and so of $\boldsymbol{q}$ (when the plate is at the BFP);
- The modified spectrum $T(\boldsymbol{q})\ F(\boldsymbol{q})$ is transformed to an amplitude distribution $g(\boldsymbol{x}_i)$ at the image plane.

The type of transform to be used depends on the geometry of the system. If both the phase distribution of the object and the phase plate geometry can be expressed in Cartesian coordinates (x, y), then Fourier transforms can be used. If both the object and the phase plate can be described in polar transverse coordinates (r, θ), then Hankel transforms are needed. The Hankel transform can be found by conversion from the Fourier transform [5], replacing x_o and y_o by ($r_o \cos\theta_o$) and ($r_o \sin\theta_o$), and q_x and q_y by ($q \cos\theta_p$) and ($q \sin\theta_p$). In either coordinate system, if the object and the plate can each be described by a single product of functions of single transverse coordinates, then the general 2D transforms reduce to products of 1D transforms.

In modelling the object phase distribution it is convenient to represent separately the magnitude of the phase variation, written here as A, and its spatial distribution, written as $\phi(x_o)$ or $\phi(r_o)$ with a maximum value of 1. The theory here is written for a weak phase object for which $A \ll 1$. Then the object exit wave f can be approximated by $[1 + iA\phi(x_o)]$ (with variation as exp i(kz−ωt) omitted). We represent the image wave by a power series in A, of the form $[1 + iA\eta(x_i) + O(A^2)]$, where a function η (as yet unknown) is used because its spatial distribution will be shown to differ from ϕ. The image amplitude is proportional to $[1 - A\,\mathrm{Im}(\eta)]$ to first order, so the intensity is proportional to $[1 - 2A\,\mathrm{Im}(\eta) + O(A^2)]$. Thus when the direct beam is present at the detector, the range of intensity due to object phase variation is proportional to A. The lens is assumed to be perfect, in the sense of having no aberrations or frequency-dependent phase shift.

3. Types of phase plate

The phase plates considered here are the Foucault (or knife-edge), Hilbert and Zernike plates. For Cartesian geometry as needed for Foucault and Hilbert plates, the object is a strip of width 2b so its phase distribution is $\phi(x_o) = 1$ for $|x_o| < b$, 0 for $|x_o| > b$. The Fourier spectrum of the function $f(x_o)$ is then $F(q) = \left[\sqrt{2\pi}\delta(q) + iA\sqrt{2/\pi}\sin qb / q\right]$.

3.1. Foucault plate

This plate is straight-edged and opaque, at a spacing defined as q_0 from the axis of the direct beam. The transmission function of the plate is $T(q) = 1$ for $q < q_0$, else = 0. The real part of the image function can be expressed using (tabulated) cosine integral functions Ci [6, 7] as $(1 - A\,I_1 / 2)$, where

$$I_1(X_i, B) = [\mathrm{Ci}\,|X_i + B| - \mathrm{Ci}\,|X_i - B|]/\pi \quad (1)$$

Here $X_i = q_0 x_i$, $B = q_0 b$ and x_i is the transverse coordinate in the image plane, measured in the same direction as q. The assumed distribution of phase is even in x_o, but the function I_1 is odd in x_i so I_1 does not model this object well. When plotted as a function of x_i or modelled by converting intensity to a monochrome scale (figure 1), I_1 gives the appearance of the derivative of the object function.

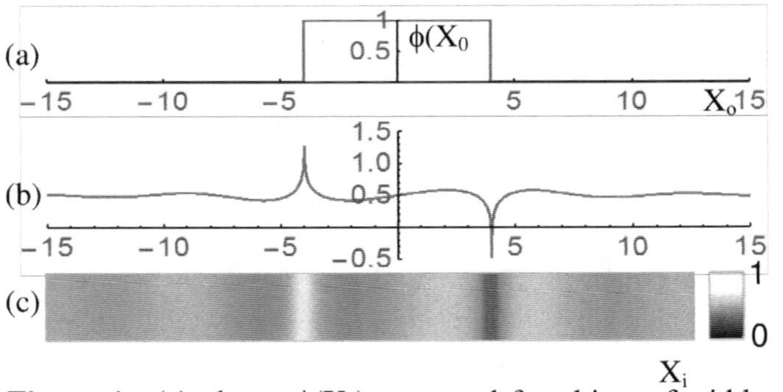

Figure 1. (a) phase $\phi(X_o)$ assumed for object of width $q_0 b = 4$; (b) plot of $0.5*(1 - I_1)$ with Foucault plate located at $q_0 = 4 / b$; (c) density plot and colour bar corresponding to (b).

3.2. Hilbert plate

The physical layout is as for the Foucault plate, but the Hilbert plate is transparent and adds a phase α relative to free space, where it is present. Its transmission function is thus

$$T_2(q/q_0) = \exp i\theta(q/q_0), \qquad \theta = \begin{cases} 0, & q/q_0 \le 1 \\ \alpha, & 1 < q/q_0 \end{cases}$$

The resulting real part of the image function is [8]

$$\mathrm{Re}(g) = 1 - A\sin\alpha/2\,[I_1 \sin\alpha/2 + I_3 \cos\alpha/2]$$

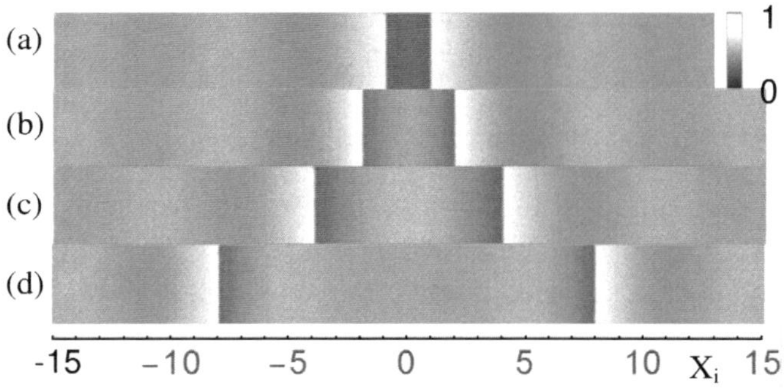

Figure 2. Variation of 0.5*(1 – I_3) with the direct beam present, for (a) B = 1; (b) B = 2; (c) B = 4; (d) B = 8. The calculated range of this function is 0.25 to 0.75 for the values of B shown, but 0 to 0.5 for B ≪ 1.

where I_1 is as defined by (1) and I_3 is defined by

$$I_3(X_i, B) = \phi(-X_i/B) - [\mathrm{Si}(X_i + B) - \mathrm{Si}(X_i - B)]/\pi$$

Here Si( ) is the sine integral function defined in [7]. The function I_3 is even in x_i (figure 2), but if α is chosen as (2nπ) to maximise cot α/2, the amplitude variation vanishes to first order. Thus an image from a Hilbert plate is likely to consist partly of function I_1 and so does not provide an accurate image of this even object.

3.3. Zernike plate with step phase transition

The rotationally symmetric object considered here is a disc of diameter 2b providing a phase change defined as $\phi(r) = 1$ for $r < b$, 0 for $r > b$. This implies a step change in the object phase at $r = b$. A Zernike plate consists of a sheet with a central circular hole, outside which the phase change relative to free space is uniform and typically π/2. Its effect can be found by Hankel transforms as indicated above and in [6, 8]. The image intensity as a function of radius can be calculated numerically with software such as Mathematica [9], which also provides conversions to area density plots.

The transition in phase at the edge of the central hole in the plate is modelled here first by a simple step at spatial frequency $q_0/2\pi$. Images of disc objects of four sizes are shown in figure 3.

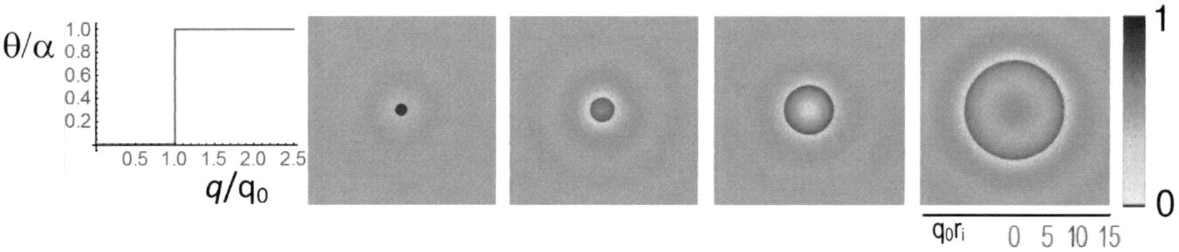

Figure 3. Step phase change introduced by a Zernike plate at the beam hole, and resulting image functions for circular phase objects of (dimensionless) radii B = q_0b = 1, 2, 4 and 8. The colour level of 0.5 at the image corresponds to zero phase added by the object.

Disc objects of size B ≤ 1 are imaged well, but the images of larger objects show radial fluctuations of amplitude that are not present in the object phase. Also the image function of figure 3, like that of figure 2, models the step change in the phase added by the object, $\phi(r)$ at $r = b$, by a change from 0.5 to 1 for B ≪ 1 but by a step from 0.25 to 0.75 for larger objects. The fluctuation of image amplitude with radius is continuous except at the step, but is depressed at radii near and on both sides of the step. The most obvious effect is a brightening of the image outside the step over a range of radii of $q_0\Delta r \sim 2$, which appears to correspond to the white outlines reported earlier [2].

3.4. Zernike plate with broadened phase transition

The radial fluctuations in image amplitude generated when there are sharp transitions in both the object phase and the phase-plate transmission resemble the fluctuations or 'ringing' that can appear when an electronic amplifier with sharply-varying frequency response is used to amplify a square wave. These effects in an amplifier can be reduced by broadening the frequency response, so the calculations for a Zernike plate were modified by broadening the range of q over which its phase

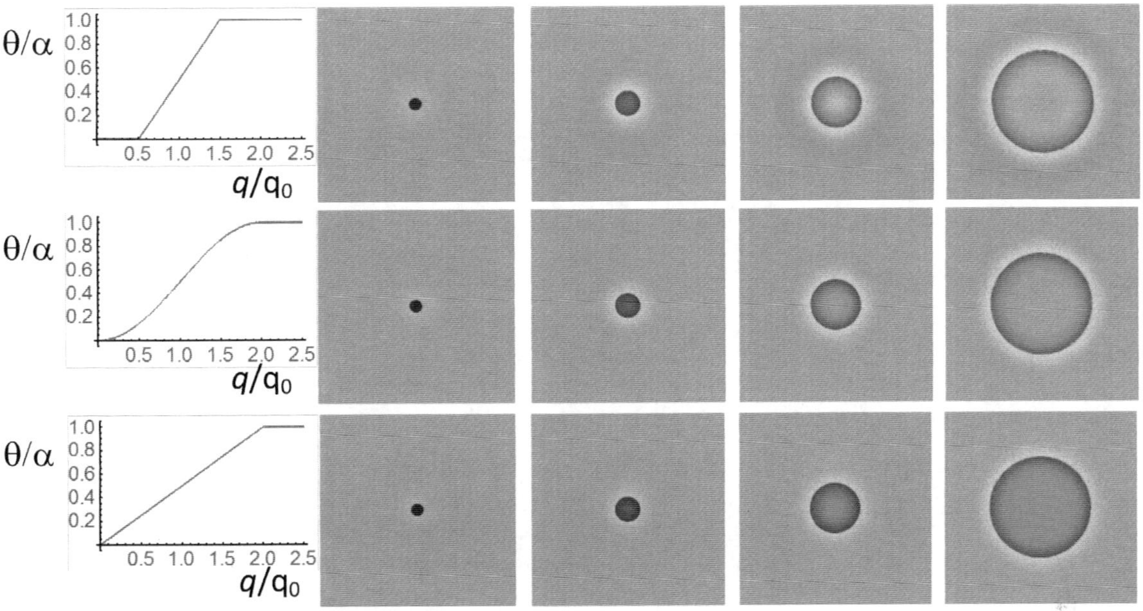

Figure 4. Plots of $\theta(q/q_0)$ and of image function for $\alpha = \pi/2$ and $B = q_0 b = 1, 2, 4$ and 8, for three distributions of phase added near the centre of a Zernike plate. The radial and colour scales for the image function are as for Fig. 3.

transition occurred. The results are shown in figure 4 for three transitions. In each case, half the total phase change occurs at $q = q_0$.

4. Conclusions

A Foucault plate or a Hilbert plate whose phase shift α is an odd multiple of π will, even with a perfect lens, produce an odd image distribution. Then the image of a rectangular weak phase object resembles a spatial derivative of the object phase. If α for a Hilbert plate has some other value, the image will include an even function together with the odd function, but as $\alpha \rightarrow 2\pi n$ the contrast at focus falls to zero.

A Zernike plate can image phase objects more accurately but, with a step change in phase at the beam hole, the intensity for objects with $B > 1$ fluctuates with radius. The fluctuation can be reduced by making the phase change occur over a broader range of radii, for instance by use of a ferromagnetic ring. The plate may add white outlines to large objects, resulting from the limitation of phase shift at radii near the direct beam radius, but is still able to provide the desired increase in phase contrast at exact focus.

References

[1] Barton B, Joos F and Schroeder R R 2008 *J Struct Biol* **164** 210-20, Fig. 2b
[2] Danev R, Glaeser R M and Nagayama K 2009 *Ultramicroscopy* **109**, 312-25 Fig. 9(b)
[3] Danev R and Nagayama K 2011 *Ultramicroscopy* **111** 1305-15 Fig. 1
[4] Born M and Wolf E 1993 (6th ed) *Principles of Optics* (Pergamon) Sec. 8.6.3
[5] Goodman J W 1996 *Introduction to Fourier Optics* (McGraw-Hill) Sec. 2.1.5
[6] Edgcombe C J 2013 *Ultramicroscopy* **136** (2014) 154–9
[7] *NIST Digital Library of Mathematical Functions* ed F W J Olver et al http://dlmf.nist.gov/6.2.ii
[8] Edgcombe C J submitted for publication
[9] Wolfram Research www.wolfram.com

4D STEM: High efficiency phase contrast imaging using a fast pixelated detector

H Yang[1], L Jones[1], H Ryll[2], M Simson[3], H Soltau[3], Y Kondo[4], R Sagawa[4], H Banba[4], I Maclaren[5], and P D Nellist[1]

1. Department of Materials, University of Oxford, Parks Rd, Oxford UK
2. PNSensor GmbH, Otto-Hahn-Ring 6, 81739 München, Germany
3. PNDetector GmbH, Sckellstraße 3, 81667 München, Germany
4. JEOL Ltd.,3-1-2 Musashino Akishima Tokyo 196-8558 Japan
5. University of Glasgow, Glasgow G12 8QQ, UK

hao.yang@materials.ox.ac.uk

Abstract. Phase contrast imaging is widely used for imaging beam sensitive and weak phase objects in electron microscopy. In this work we demonstrate the achievement of high efficient phase contrast imaging in STEM using the pnCCD, a fast direct electron pixelated detector, which records the diffraction patterns at every probe position with a speed of 1000 to 4000 frames per second, forming a 4D STEM dataset simultaneously with the incoherent Z-contrast imaging. Ptychographic phase reconstruction has been applied and the obtained complex transmission function reveals the phase of the specimen. The results using GaN and Ti, Nd-doped $BiFeO_3$ show that this imaging mode is especially powerful for imaging light elements in the presence of much heavier elements.

1. Introduction

Phase contrast imaging using elastic electron scattering has been shown to provide the highest dose efficiency for imaging weak phase objects for a given amount of radiation damage, as compared to inelastic electron scattering, as well as to X-ray and neutron scattering[1]. Therefore phase contrast imaging in the conventional transmission electron microscope (CTEM) continues to play a critical role in many biological studies.

There has been growing interests in developing phase contrast imaging modes in the scanning transmission electron microscopy (STEM), because of its flexibility in allowing innovations in detector geometry without modifying the main optics of the electron column and the ability to simultaneously record incoherent and analytical signals. For example, annular bright field (ABF) [2-4] and differential phase contrast (DPC) [5,6] have been demonstrated as being sensitive to light elements and electrical/magnetic fields. However, recent studies have shown that bright field (BF) and ABF are not efficient for imaging weak phase objects because interferences between diffracted and direct beams carrying positive and negative phases cancel under zero aberrations using any centrosymmetric detector geometry[7,8]. The greatest signal is obtained if the regions of constructive and destructive interferences are covered by separate detectors[9].

This can be achieved by using fast pixelated detectors that record the CBED pattern at every probe position forming a 4-dimensional dataset. The phase can be extracted through post-processing using a method called ptychography[10], which makes use of the redundancy in the 4D dataset. Various ptychographic reconstruction methods have been developed in both electron and X-ray microscopy to retrieve the phase through either iterative[11,12] or direct inversion approaches[13–15]. Partially due to the practical difficulties in recording the 4D dataset due to limited camera speed, the most common practice in literature was to use a defocused probe[16] to increase the illuminated area, thus reducing the number of probe positions needed to ensure overlapping between the neighboring probe illuminating areas in order to solve the ambiguity in phase. However, with recent advances in detector technology, which have massively improved the camera speed, 4D STEM can be recorded with a large number of probe positions and a sub-Ångström focused probe that makes full use of the aberration corrector. In this way incoherent Z-contrast imaging with a high angle annular dark field (HAADF) detector and even analytical signals can be obtained simultaneously with the diffraction patterns that yield the reconstructed phase.

Here in this work we show that making use of a fast pixelated detector installed on an aberration corrected STEM, 4D datasets recorded using an aberration corrected electron probe combined with ptychography offers a new phase contrast imaging mode in STEM and opens opportunities for imaging weak phase objects and detecting light elements with high efficiency.

2. Experimental methods

The experiments were performed using a pnCCD (S)TEM camera, a direct electron radiation-hard pixelated detector from PNDetector GmbH, mounted on a JEOL ARM200CF aberration corrected microscope. This camera delivers a readout speed of 1000 full frames per second (fps) routinely[17], and up to 20,000 fps through binning/windowing. For a typical STEM scanning consisting of 256x256 probe positions, a 4D dataset consisting of 264x264 pixel diffraction patterns for each probe position can be recorded in less than 70 seconds. Figure 1 shows single frames of the convergent beam electron diffraction (CBED) patterns recorded at different detector binning levels and frame times, and single electron events can be seen clearly in the dark field regions. Synthetic BF, ABF and DPC images can also be obtained from the diffraction patterns, with the full flexibility to tune the collection angles in ABF and partition direction of the DPC segments. Z-contrast HAADF images were usually recorded simultaneously using an annular dark field (ADF) detector.

3. Results and discussion

The ptychographic reconstruction approach applied to the 4D datasets used in this work follows Rodenburg *et al.*[14] The reader is referred to the previous works[14,7] for the details of the reconstruction procedure. By taking the Fourier transform of the 4D dataset with respect to probe positions, interferences between the direct beam and the diffracted beams of specific diffraction angles can be isolated in the detector plane, and the phase information can be extracted from the regions of disc-overlapping between the direct and diffracted beams. An experimental example of the overlapping discs from an Au nanoparticle sample at a single spatial frequency is shown in Figure 2. The two diffracted discs overlap with the direct beam to form single overlap regions, and all three discs overlap to form a double overlap region in the middle. Under the weak-phase approximation, the two sides of the single overlap regions are "flat" and π out of phase, and the phases in the double overlap completely cancel. Therefore the complex quantity can be integrated from one side of the disc-overlapping regions at each spatial frequency, and the phase can be reconstructed by calculating the inverse Fourier transform of the integrated phase with respect to probe positions.

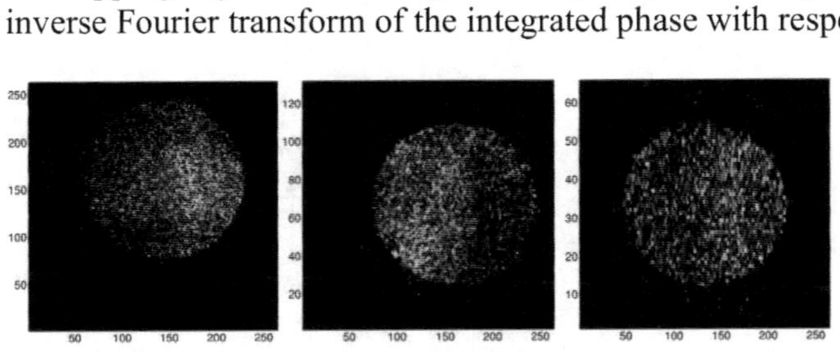

Figure 1. Examples of individual frames of CBED patterns consisting of primarily the bright field disk with three different binning and speed: (a) 1000 fps without binning, (b) 2000fps with 2-fold binning and (c) 4000 fps with 4-fold binning.

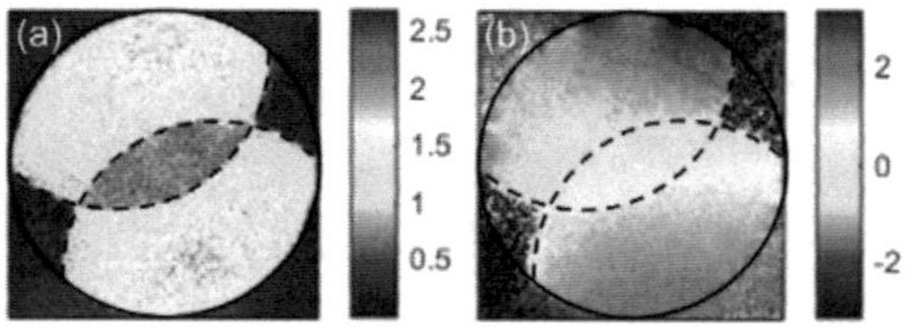

Figure 2. An experiment example of the (a) modulus and (b) phase of the interference between the direct beam (solid circle) and two diffracted beams (dashed circles) in the detector plane, showing single and double disc-overlapping regions. The colour-bar in (b) is in unit of radian.

Experiments have been applied to an Au nanoparticle on a carbon support. The simultaneous HAADF, ABF, and the modulus and phase of the phase reconstruction from an Au nanoparticle with five-fold twinning have been obtained and shown in Figure 3a-d). The ptychographic phase reconstruction shows good contrast in both the modulus and phase across thick and thin regions of the nanoparticle. In comparison, the synthetic ABF image shows decreasing contrast towards the edge of the Au nanoparticle, suggesting that ABF, as a nonlinear imaging mode, is not very efficient when the specimen is very thin, and relies on phase shift from lens aberrations and defocus to form the image contrast for thin specimens that obey a weak phase object approximation[3]. The improved contrast in ABF with increasing thickness can be ascribed to the multiple scattering channelling processes[18]. Furthermore, the ptychographic reconstruction shows a clearer texture in the carbon support region than in the ABF, indicating its high sensitivity to weakly scattered objects. Moreover, based on the reconstructed phase in unit of radians, the projected potential can be directly calculated using the interaction constant, and the atomic electric field could also be derived by calculating the first derivative of the potential[19].

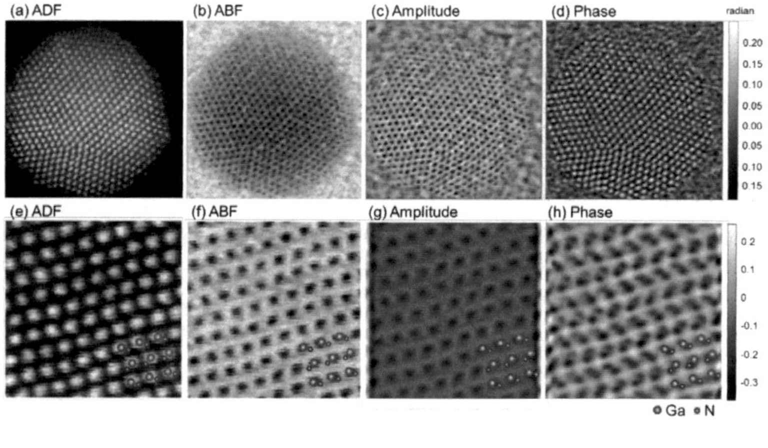

Figure 3. Simultaneous ADF, synthetic ABF, and the reconstructed amplitude and phase of (a-d) an Au nanoparticle and (e-h) GaN bulk lattice viewed along $[2\bar{1}\bar{1}0]$. The probe forming aperture for cases is about 14.4 mrad, and the synthetic ABF has a collection angle of 7.2-14.4 mrad.

The results of ptychographic phase reconstruction of GaN viewed along $[2\bar{1}\bar{1}0]$ demonstrates the sensitivity to imaging light elements, as shown in Figure 3e-h. N columns are not clearly visible in the ABF image (Figure 3f) using a 14mrad semi-angle convergence aperture with a theoretical resolution of about 0.9 Å. In contrast, the Ga-N dumbbells are clearly resolved in the reconstructed complex transmission function. The N columns give their clearest contrast in the phase image (Figure 3h). And in the modulus the major peak corresponds to the heavy Ga column because the modulus contrast comes primarily from absorption (Figure 3g).

Ptychographic phase reconstruction has also been applied to charged antiphase domain boundaries in Nd, Ti doped $BiFeO_3$[20,21]. Oxygen columns are clearly resolved in both the bulk and the boundary (Figure 4), which are not seen in the HAADF image, and are barely visible in the ABF image. The Fe/Ti columns are strong features in the reconstructed phase, as might be expected. On the other hand, the Bi columns show broad but rather low intensity diffused features in the reconstructed phase. This is almost certainly caused by strong phase shifts on these high-Z columns, leading in the direction of a contrast inversion, as is commonly observed in phase contrast high resolution TEM (HRTEM).

Nevertheless, the region is still thin enough that all atoms are imaged in positive contrast using this ptychographic reconstruction method under the weak phase approximation.

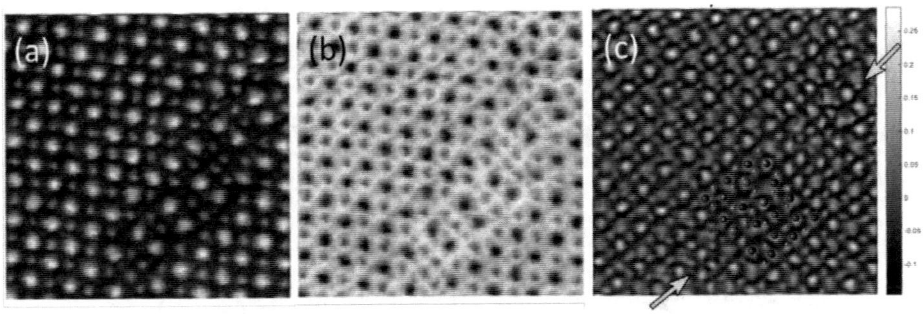

Figure 4. HAADF, synthetic ABF and the reconstructed phase of an antiphase boundary in Ti, Nd doped $BiFeO_3$. It should be noted that the oxygen positions which are invisible in the HAADF image appear in the phase image as small yellow peaks. The larger red peaks are B-site Fe columns, the blue peaks are B-site Ti columns, and the bright purple peaks are A-site (Bi/Nd) columns.

4. Conclusion

In conclusion we have shown that using a pnCCD pixelated detector, a highly efficient phase contrast imaging mode may be achieved by applying ptychography to the 4D STEM datasets, which opens various fields of applications for imaging weak phase objects and light elements, which are complementary to the well-established incoherent Z-contrast imaging in STEM.

The authors acknowledge funding from the EPSRC (grant numbers EP/K032518/1, EP/K040375/1 EP/M009963/1 and EP/M010708/1) and the EU FP7 Grant Agreement 312483 (ESTEEM2).

References

[1] Henderson, R. *Q. Rev. Biophys.* **28,** 171–193 (1995)
[2] Rose, H. *Ultramicroscopy* **2,** 251–267 (1976)
[3] Ishikawa, R. *et al. Nat. Mater.* **10,** 278–81 (2011)
[4] Okunishi, E. *et al. Microsc. Microanal.* **15**, 164-165 (2009)
[5] Dekkers, N. H. & Lang, H. D. *Optik.* **41,** 452–456 (1974)
[6] Shibata, N. *et al. Nat. Phys.* **8,** 611–615 (2012)
[7] Pennycook, T. J. *et al. Ultramicroscopy* **151,** 160–167 (2015)
[8] Yang, H. *et al. Ultramicroscopy* **151,** 232–239 (2015)
[9] Rose, H. *Ultramicroscopy* **2,** 251–67 (1977)
[10] Hoppe, W. *Acta Crystallogr. A* **25**, 495-501 (1969)
[11] Maiden, A. M.*et al. Ultramicroscopy* **109,** 1256–62 (2009)
[12] Thibault, P. *et al. Science* **321,** 379–82 (2008)
[13] Rodenburg, J. M. *et al. Philo. Trans.Roy. Soc. A* **339,** 521–553 (1992)
[14] Rodenburg, J. M. *et al. Ultramicroscopy* **48,** 304–314 (1993)
[15] D'Alfonso, A. J. *et al. Phys. Rev. B.* **89,** 064101 (2014)
[16] Humphry, M. J. *et al. Nat. Commun.* **3,** 730 (2012)
[17] Ryll, H. *et al. Microsc. Microanal.* **20,** 1122–1123 (2014)
[18] Findlay, S. *et al. Ultramicroscopy* **110**, 903-923 (2010)
[19] Cowley, J. M. *Electron Diffraction Techniques. Vol. 1.* Oxford University Press (1992)
[20] MacLaren, I. *et al. APL Mater.* **1,** 021102 (2013)
[21] MacLaren, I. *et al. APL Mater.* **2,** 066106 (2014)

Resolution Improvement in Aberration-Corrected Low-Voltage TEM with Monochromator at 60 kV

S Morishita[1,3], M Mukai[1,3], T Sasaki[1,3], K Suenaga[2,3], H Sawada[1,3]

[1] JEOL Ltd. 1-2 Musashino 3-Chome, Akishima, Tokyo 196-8558 Japan
[2] National Institute of Advanced Industrial Science and Technology, 1-1-1 Higashi, Tsukuba, Ibaraki 305-8565 Japan
[3] Research acceleration program, Japan Science and Technology Agency, K's Gobancho, 7, Gobancho, Chiyoda-ku, Tokyo, 102-0076 Japan

E-mail: shmorish@jeol.co.jp

Abstract. We have developed a low-voltage electron microscope equipped with a monochromator and Delta-type Cs correctors, which shows atomic resolution at accelerating voltages of 60, 30 and 15 kV. In theory, resolution of TEM images at 60 kV is severely affected by chromatic aberration, which is proven by our calculations of contrast transfer functions and multi-slice image simulation taking chromatic aberration into account with experimental conditions. Experimentally, TEM images of gold nano-particles were observed with non-monochromated and monochromated electron sources at 60 kV. Detectable spatial frequency in the image with the monochromated source was higher than that with non-monochromated source. We have demonstrated that the TEM image resolution at the low-voltage is improved by using a monochromated electron source, which reduce the energy spread of the electron source.

1. Introduction

Low-voltage transmission electron microscope (LVTEM) becomes essential tool to study nano-structure and chemical analysis of carbon related materials, since the LVTEM enables us to observe them with small radiation damage. In theory, the resolution in LVTEM is limited by diffraction limit ($d_d = 0.61\ \lambda / \alpha$, where λ and α are wavelength of electrons and acceptance angle for the imaging). The wavelength of electrons at 60 kV is 4.87 pm, which is approximately double of that for 200 kV, resulting in double diffraction limit with identical aperture angle. To improve the resolution at low accelerating voltage, reduction of the diffraction limit is mandatory. Therefore, we have developed a Delta-type aberration corrector that is composed of three dodecapoles [1]. It can correct fifth-order aberrations including six-fold astigmatism as well as third-order spherical aberration. With the corrector in probe forming system, atomic spacing of carbon-to-carbon in a graphene was clearly resolved in STEM images at 30 kV [2]. The Delta corrector can be applied to image forming lens system. Diffraction limit can be decreased by the large acceptance angle for the imaging realized by the corrector. Though we applied the Delta corrector for image forming lenses experimentally, the resolution of TEM images was not enough to detect the atomic structure of mono-layered materials [3]. It was concluded that the result was caused by a chromatic aberration, because TEM resolution is affected by defocus spread more severely than the resolution of a dark field STEM. Defocus spread caused by the chromatic aberration is defined as C_c dE / E_0 where C_c, dE and E_0 are chromatic aberration coefficient, energy spread of electron source and primary electron energy. The defocus spread becomes larger when E_0 becomes smaller with constant dE. To improve the resolution in TEM images at lower accelerating voltages, correction of chromatic aberration C_c of the image forming lenses or reduction of the energy spread dE are required.

2. Experimental methods

In the present study, a newly developed LVTEM equipped with a monochromator and a Delta corrector for image forming lens systems was used. The microscope can work at accelerating voltage of 60, 30 and 15 kV. The accelerating voltage of 60 kV was used in this work. The monochromator consists of two Wien filters and an energy selection slit [4]. The energy width of the electron source can be changed by selecting slit width. The minimum energy width of the electron source was measured to be less than 25 meV with exposure time of 0.5 seconds. The coefficients of third-, fourth-, and fifth-order aberrations have been measured to be less than 1 μm, 30 μm, 0.1 mm, respectively. The fifth-order aberration of 0.1 mm induces $\pi/4$ phase shift in 57 mrad, which corresponds to the spatial frequency of higher than $(0.09 \text{ nm})^{-1}$. The measured chromatic aberration coefficient at 60 kV for the image forming system was 0.75 mm.

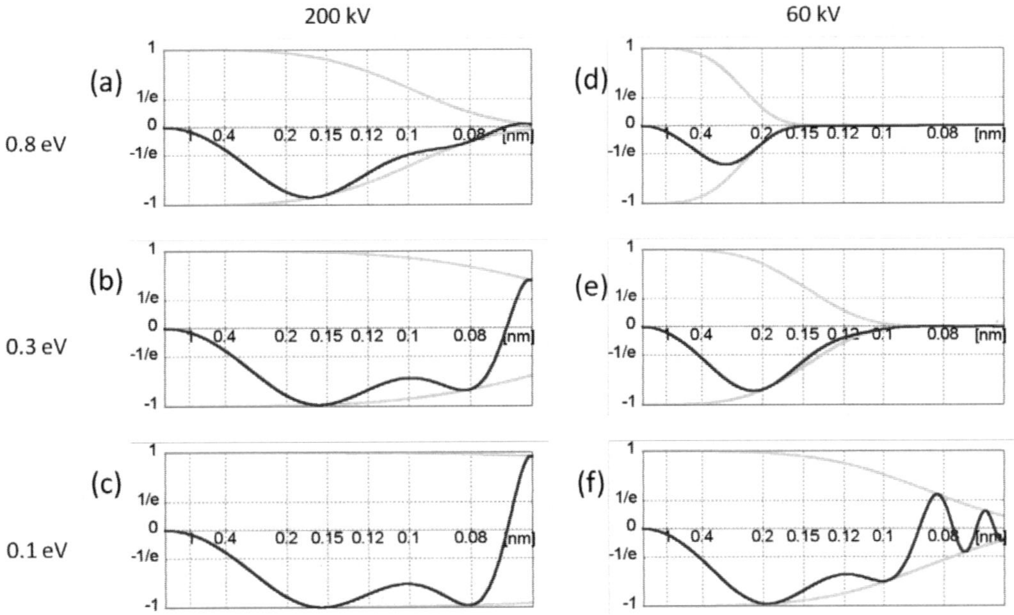

Figure 1. Calculated contrast transfer functions (blue) and envelope functions (green) at 200 kV ((a), (b) and (c)) and 60 kV ((d), (e) and (f)). Energy width of (a) and (d) is 0.8 eV, and that of (b) and (e) is 0.3 eV, and that of (c) and (f) is 0.1 eV. Chromatic aberration coefficient is 0.75 mm. Third-order spherical aberration coefficient and defocus are 10 μm and -6.1 nm at 200 kV and 3 μm and -4.7 nm at 60 kV, respectively.

3. Simulations

To evaluate the effect of the energy spread of an electron source on TEM images at lower accelerating voltage, we have calculated contrast transfer function (CTF) at two accelerating voltages. Figure 1 shows the CTFs at 200 and 60 kV with several energy spreads of the electron source. Chromatic aberration coefficient of 0.75 mm is used for the calculation. Here, an indicator of the spatial frequency for the information limit is defined as the frequency when intensity of the envelope function reduces to be 1/e (= 0.368). In the case of 200 kV, information limit of the envelope function is higher than $(0.1\ \text{nm})^{-1}$ even if the energy width is 0.8 eV, which is the typical energy width of a Schottkey source. On the other hand, in 60 kV, it is difficult to achieve $(0.1\ \text{nm})^{-1}$ with the energy width of even 0.3 eV which corresponds to the typical energy spread of a cold field-emission source. To achieve the information limit better than $(0.1\ \text{nm})^{-1}$ at 60 kV, monochromated source is required. This indicates that the reduction of the energy spread is important for improving the information limit at 60 kV.

Figure 2 shows simulated TEM images of a gold nano-particle from <111> direction at 60 kV by the multi-slice method. The effect of the defocus spread is included in the simulation by multiplying the Fourier transform of the exit wave by the envelope function due to the defocus spread. When the energy width is 0.8 eV, the contrast of atomic columns in the particle are blurred. By reducing the energy width to be 0.1 eV, the atomic columns are clearly resolved in the image. The simulation indicates the resolution enhancement using a monochromated source in TEM images at 60 kV.

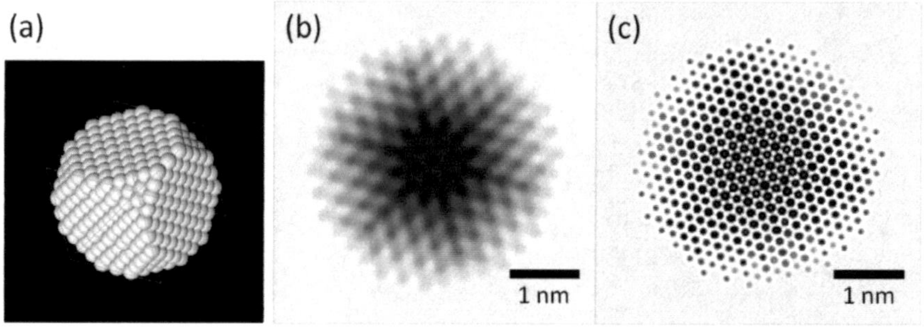

Figure 2. Atomic model (a) and simulated TEM images of a gold nano-particle from <111> with the energy width of (b) 0.8 eV, and (c) 0.1 eV. Chromatic aberration coefficient of 0.75 mm and defocus of - 2.0 nm are used in the simulation.

4. Results & Discussion

Figures 3 show TEM images of gold particles and their power spectra of the Fourier transforms taken at 60 kV with the different energy spreads of the electrons. Non-monochromated electron source without energy slit and monochromated source with the slit width of 1.3 μm were chosen for the imaging conditions. The energy width of these electron sources were 0.77 eV and 0.10 eV, respectively. The atomic columns of the gold nano-particles were not clearly observed in the image with the non-monochromated source [Fig. 3(a)], though lattice fringes were detected in the image. On the other hand, the atomic columns of the gold nano-particles were detectable in the image with the monochromated source [Fig. 3(b)]. The power spectrum of the image from the non-monochromated source shows that the continuous bright contrast, which is originated by structural information of the amorphous carbon film, exhibits up to around $(0.2\ \text{nm})^{-1}$. The contrast in the power spectrum with the monochromated source extends to around $(0.1\ \text{nm})^{-1}$. Spots around $(0.1\ \text{nm})^{-1}$ from the gold-crystalline structure were detected in the image with monochromated source. Although these power spectra cannot be simply used for evaluating the resolution of the TEM images quantitatively because of non-linear contrast [5], these images and power spectra apparently show that the resolution of the TEM images is improved by using monochromated electron source.

The resolution of TEM images can be affected by both envelope functions due to spatial coherence and temporal coherence of the electron source, generally. Our monochromator produces achromatic and stigmatic illumination beam after the second Wien filter. Since a small cross-over for illumination makes a small illumination angle, the illumination angle for imaging is sufficiently small. And the aberrations up to sixth order were corrected by the image-forming aberration corrector. Thus, the effect of the envelope function due to the spatial coherence, which is caused by the illumination angle and the geometrical aberration, is sufficiently small. Therefore, the effect of the envelope function due to the spatial coherence can be neglected in the TEM images in Fig. 3. The resolution was dominantly affected by the envelope function due to temporal coherence in this observation.

We experimentally demonstrated that the resolution of the aberration-corrected TEM image was enhanced by the monochromated electron source. The experimental result indicates that the energy spread of the electron source is a dominant factor for information transfer at low accelerating of 60 kV [6], whereas the defocus spread from fluctuations of accelerating voltage and lens currents can affect the resolution in TEM.

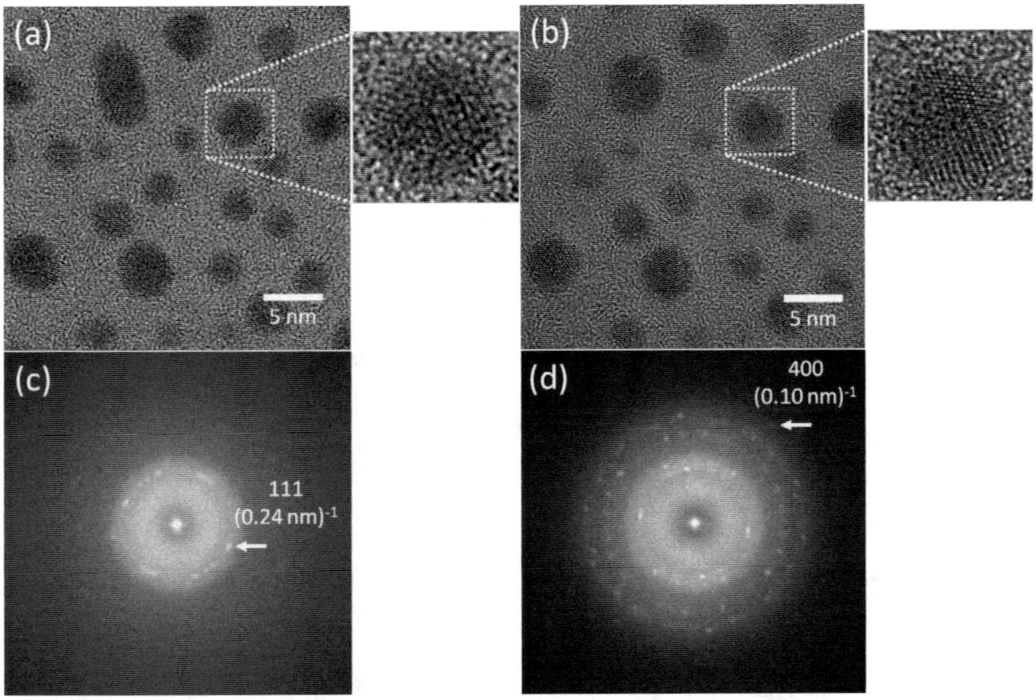

Figure 3. TEM images of gold nano-particles taken at 60 kV with (a) non-monochromated source and (b) monochromated source. (c) and (d) are power spectra of the Fourier transform of (a) and (b), respectively.

5. Conclusion

We have developed the aberration-corrected LVTEM equipped with the monochromator, which shows the atomic resolution at 60 kV. The CTFs and the simulated TEM images show that the information limit can be improved by reducing the energy spread of the electron source. The TEM images of the gold nano-particles with non-monochromated and monochromated electron source, taken at 60 kV, showed that narrower energy spread of the electron source improved the resolution of TEM images.

Acknowledgments

This work was supported by the JST under the Research Acceleration Program (2012–2016).

References

[1] Sawada H, Sasaki T, Hosokawa F, Yuasa S, Terao M, Kawazoe M, Nakamichi T, Kaneyama T, Kondo Y, Kimoto K and Suenaga K 2010 *Ultramicroscopy* **110** 958-61.
[2] Sawada H, Sasaki T, Hosokawa F and Suenaga K 2015 *Phys. Rev. Let.* **114** 166102.
[3] Sasaki T, Sawada H, Hosokawa F, Sato Y and Suenaga K 2014 *Ultramicroscopy* **145** 50-55.
[4] Mukai M, Kim J S, Omoto K, Sawada H, Kimura A, Ikeda A, Zhou J, Kaneyama T, Young N P, Warner J H, Nellist P D and Kirkland A I 2014 *Ultramicroscopy* **140** 37-43.
[5] O'Keefe M A 1994 *Ultramicroscopy* **47** 282-297.
[6] Warner J H, Margine E R, Mukai M, Robertson A W, Giustino F and Kirkland A I 2012 *Science* **337** 209-212.

Quantitative annular dark field scanning transmission electron microscopy for nanoparticle atom-counting: What are the limits?

A De Backer[1], A De wael[1], J Gonnissen[1], G T Martinez[1], A Béché[1], K E MacArthur[2], L Jones[2], P D Nellist[2] and S Van Aert [1]

[1] Electron Microscopy for Materials Science (EMAT), University of Antwerp, Groenenborgerlaan 171, B-2020 Antwerp, Belgium

[2] Department of Materials, Uni versity of Oxford, 16 Parks Road, Oxford OX1 3PH, UK

E-mail: annick.debacker@uantwerpen.be

Abstract. Quantitative atomic resolution annular dark field scanning transmission electron microscopy (ADF STEM) has become a powerful technique for nanoparticle atom-counting. However, a lot of nanoparticles provide a severe characterisation challenge because of their limited size and beam sensitivity. Therefore, quantitative ADF STEM may greatly benefit from statistical detection theory in order to optimise the instrumental microscope settings such that the incoming electron dose can be kept as low as possible whilst still retaining single-atom precision. The principles of detection theory are used to quantify the probability of error for atom-counting. This enables us to decide between different image performance measures and to optimise the experimental detector settings for atom-counting in ADF STEM in an objective manner. To demonstrate this, ADF STEM imaging of an industrial catalyst has been conducted using the near-optimal detector settings. For this experiment, we discussed the limits for atom-counting diagnosed by combining a thorough statistical method and detailed image simulations.

1. Introduction

The observed intensity in atomic resolution STEM images is highly sensitive to the number of projected atoms in each atomic column. Therefore, quantitative ADF STEM has become a popular technique for nanoparticle atom-counting. In existing atom-counting methods the number of atoms is estimated from an ADF STEM image by a comparison with detailed image simulations or by using advanced statistical methods [1–4]. By analysing the total fraction of scattered electrons from each atomic column, the so-called scattering cross-section, it has been shown that the number of atoms in a single-element column can be counted with single-atom sensitivity on model systems, which are relatively stable under the incoming electron beam [2–5]. Here we discuss how we can extend this technique to count the number or atoms in more challenging nanostructures.

2. Statistical detection theory

In order to keep the incoming electron dose as low as possible but still retain single-atom precision, statistical detection theory can be used to optimise the instrumental microscope settings [6]. Similar to the use of statistical parameter estimation theory to quantify the precision

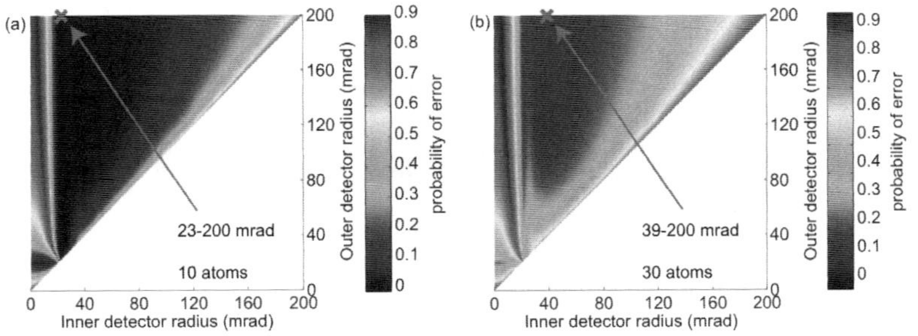

Figure 1. Probability of error as a function of inner and outer detector angle for Pt along the [110] zone axis for an electron dose of 10^4 $e^-/\text{Å}^2$. (a) Probability of error for counting up to 10 atoms. (b) Probability of error for counting up to 30 atoms.

with which atom column can be located [7–10], the principles of detection theory allow one to quantify the probability of error for atom-counting [11]. This probability of error has been used to compare different image performance measures that can be used for atom-counting. Such analysis suggests that the scattering cross-sections perform equally well as the image intensities on a pixel by pixel basis and prove more robust than peak intensities. Moreover, based on an expression for the probability of error, it can be shown that for very thin objects low angle ADF STEM is optimal and that for thicker objects the optimal inner angle increases as illustrated in Fig. 1 [11].

3. Atom-counting on a Pt/Ir particle

Next, atom-counting will be demonstrated on an industrial bimetallic catalyst. Four images of the same Pt/Ir particle were recorded at different magnifications and electron doses under near-optimal detector settings (i.e. 35-190 mrad) at the QuAntEM, a double corrected FEI Titan[3] working at 300 kV with a 20.2 mrad semi-convergence angle. Because of their beam sensitivity and limited size, these particles provide a severe characterisation challenge, even under the optimal detector settings [12]. By combining simulation- and statistics-based methods, being completely independent of each other, the effect of dose and limited particle size on the reliability of atom-counting can be determined. In order to allow comparison with simulations, the images were normalised with respect to the incoming electron beam intensity [13]. Next, using statistical parameter estimation theory, the scattering cross-sections were quantified atomic column - by - atomic column. For this Pt/Ir particle, both atomic species ($Z_{\text{Ir}}=77$ and $Z_{\text{Pt}}=78$) have been treated in the same way, since the scattering cross-sections of Pt cannot be distinguished from the scattering cross-sections of Ir up to 30 atoms. An example analysis for the image recorded at the highest magnification (pixel size of 0.12 Å) and electron dose (5.2×10^5 $e^-/\text{Å}^2$) is illustrated in Fig. 2; the scattering cross-sections are visualised in a histogram. The number of significant components and their scattering cross-sections were determined by evaluating the so-called integrated classification likelihood (ICL) criterion in combination with Gaussian mixture model estimation. These results allow us to quantify the number of atoms in each atomic column.
In Fig. 3, the experimental scattering cross-sections, i.e. the components of the Gaussian mixture model, resulting from the counting analyses are compared with the scattering cross-sections resulting from simulated images using STEMsim [14]. For image D (shown in Fig. 2(a)) with the highest magnification and electron dose, an excellent match was found. However, analysing images of lower magnification (less finely sampled image) and/or lower electron dose worsens the match with simulation. The same effect is observed when analysing a subsampled image, i.e. composed of every second pixel of image D, whereas the sampling has no effect on the reliability

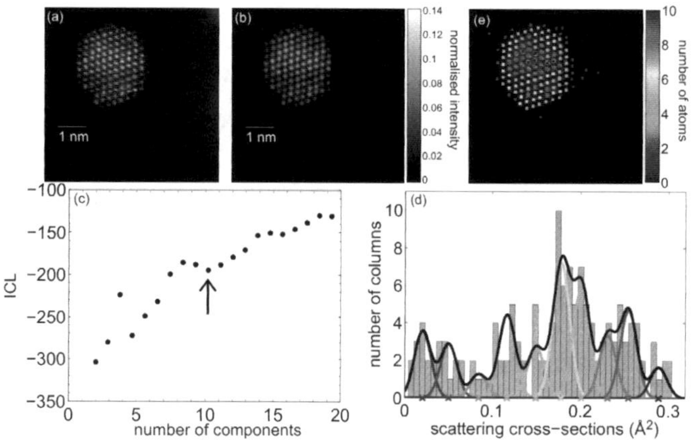

Figure 2. Illustration of the atom-counting procedure. (a) Experimental ADF STEM image. (b) Refined parametrised imaging model. (c) Evaluation of ICL as a function of the number of components. (d) Histogram of estimated scattering cross-sections together with the estimated Gaussian mixture model. (e) Quantification of the number of atoms in each atomic column.

as shown by the binned version of image D, i.e. by averaging the pixels. In the subsampled image, the lower dose of images A and B is mimicked. This leads to less precise measurements of the scattering cross-sections resulting in insufficient statistics for the determination of the number of components. However, when enhancing the statistics by combining the values of the scattering cross-sections of the four images collectively, the experimental values again match with simulated values. The study demonstrates that while the incoming electron dose is often lowered in order to reduce sample damage, the detected dose should be sufficiently large to retain single-atom precision, even when the image has been recorded under the optimal detector settings.

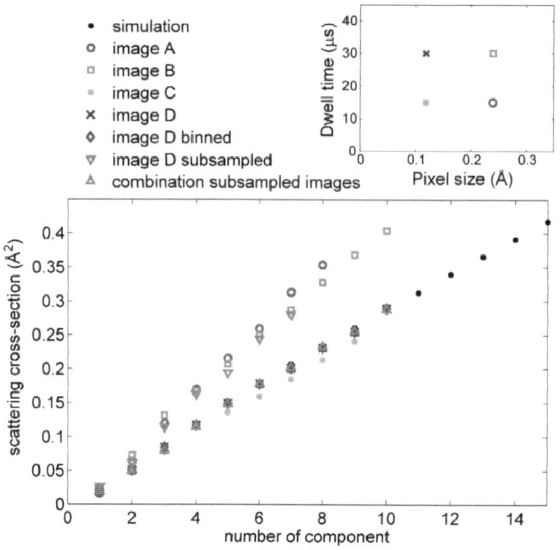

Figure 3. Comparison of experimental and simulated scattering cross-sections. The upper right graph shows the specific pixel size and dwell time for the individual images of the Pt/Ir particle.

4. Conclusions

In conclusion, using statistical detection theory, the detector settings can be optimised which becomes extremely important when looking at beam sensitive systems. A lower inner detector angle is suggested for thin objects in order to maximise the detected dose. This detector setting has been used for an atom-counting analysis of an industrial catalyst. It has been shown that even when employing optimal detector settings, a minimum level of electron dose is required to retain single-atom precision. The reliability of the atom-counting results could be studied by using a combination of advanced statistical methods and detailed image simulations [14].

Acknowledgments

The authors acknowledge financial support from the Research Foundation Flanders (FWO, Belgium) through project funding (G.0368.15N, G.0369.15N, and G.0374.15N) and a PhD research grant to A De Backer. The research leading to these results has received funding from the European Union Seventh Framework Programme under Grant Agreement 312483 - ESTEEM2 (Integrated Infrastructure Initiative-I3), ERC Starting Grant 278510 Vortex, and the UK Engineering and Physical Sciences Research Council (EP/K032518/1). The authors acknowledge Johnson-Matthey for providing the sample and PhD funding to K E MacArthur. A Rosenauer is acknowledged for providing the STEMsim program.

References

[1] LeBeau J M, Findlay S D, Allen L J and Stemmer S 2010 *Nano Letters* **10** 4405–4408

[2] Van Aert S, Batenburg K J, Rossell M D, Erni R and Van Tendeloo G 2011 *Nature* **470** 374–377

[3] Van Aert S, De Backer A, Martinez G T, Goris B, Bals S and Van Tendeloo G 2013 *Physical Review B* **87**

[4] De Backer A, Martinez G T, Rosenauer A and Van Aert S 2013 *Ultramicroscopy* **134** 23–33

[5] E H, MacArthur K E, Pennycook T J, Okunishi E, D'Alfonso A J, Lugg N R, Allen L J and Nellist P D 2013 *Ultramicroscopy* **133** 109–119

[6] Kay S M 2009 *Fundamentals of Statistical Signal Processing: Volume II Detection Theory* (New Jersey: Prentice-Hall, Inc.)

[7] den Dekker A J, Van Aert S, Van Dyck D, van den Bos A and Geuens P 2001 *Ultramicroscopy* **89** 275–290

[8] Van Aert S, den Dekker A J, Van Dyck D and van den Bos A 2002 *Journal of Structural Biology* **138** 21–33

[9] De Backer A, Van Aert S and Van Dyck D 2011 *Ultramicroscopy* **111** 1475–1482

[10] den Dekker A J, Gonnissen J, De Backer A, Sijbers J and Van Aert S 2013 *Ultramicroscopy* **134** 34–43

[11] De Backer A, De wael A, Gonnissen J and Van Aert S 2015 *Ultramicroscopy* **151** 46–55

[12] De Backer A, Martinez G T, MacArthur K E, Jones L, Béché A, Nellist P D and Van Aert S 2015 *Ultramicroscopy* **151** 56–61

[13] Rosenauer A, Gries K, Müller K, Pretorius A, Schowalter M, Avramescu A, Engl K and Lutgen S 2009 *Ultramicroscopy* **109** 1171–1182

[14] Rosenauer A and Schowalter M 2008 Stemsim - a new software tool for simulation of STEM HAADF Z-contrast imaging *Microscopy of Semiconducting Materials 2007* (*Springer Proceedings in Physics* vol 120) ed Cullis A and Midgley P (Springer Netherlands) pp 170–172

Quantification of a Heterogeneous Ruthenium Catalyst on Carbon-black using ADF Imaging.

A M Varambhia[1], L Jones[1], P D Nellist[1], S Lozano-Perez[1], D Ozkaya[2]
[1]Department of Materials, University of Oxford, Parks Road, OX1 3PH, UK.
[2]Johnson Matthey Technical Centre, Sonning Common, Reading, RG4 9NH, UK.

Email: aakash.varambhia@materials.ox.ac.uk

Abstract. Using recent developments in ADF quantification techniques, we utilise a rapid atom counting procedure to document the evolution of a heterogeneous ruthenium catalyst on carbon black. Selected areas of the ruthenium catalyst were imaged for approximately 15 minutes. The imaged areas show that the Ru wets the support forming a thin amorphous layer that transforms to a crystalline layer under beam irradiation.

1. Introduction.

Ruthenium catalysts show promising results for many catalytic processes, however, their crystal structure (hexagonal closed packed) prevented them being considered as potential candidates for many reactions. With the recent discovery of fcc (face centred cubic) ruthenium nanoparticles[1], interest in ruthenium has picked up again. The fcc structure has been attributed to a substantial catalytic activity enhancement in CO oxidation reactions when compared to the more common hcp ruthenium nanoparticles[1].

Here we study a ruthenium on carbon catalyst using annular dark-field (ADF) scanning transmission electron microscopy (STEM). ADF STEM is particularly suited to the study of heterogeneous catalysts because of its strong dependence on atomic number and lack of contrast reversals with thickness and defocus. Particle stability under the electron beam remains a challenge with this approach, however.

In this paper, using recently developed quantification techniques, we present results of a pure heterogeneous ruthenium catalyst supported on carbon-black, changing structurally, as it was imaged under the electron beam. The images were normalised with the appropriate parameters and atomic column counts were obtained for the images.

2. Experimental and normalisation parameters.

The nanoparticles were imaged using a JEOL ARM 200CF EM at NTNU (Trondheim, Norway) was done at 200kV with 69.2mrad and 137.53mrad inner and outer detector angles respectively. Two areas of the catalyst were imaged (Figure 1) and labelled as nodule and ring, with their respective analysis regions highlighted by the yellow box.

33 and 44 frames were acquired for the nodule and ring regions respectively. The before and after labels indicate a change from amorphous to crystalline phase from the beginning to the end of the imaging period. In order to quantify each frame reliably there are certain normalisation parameters that have to be addressed.

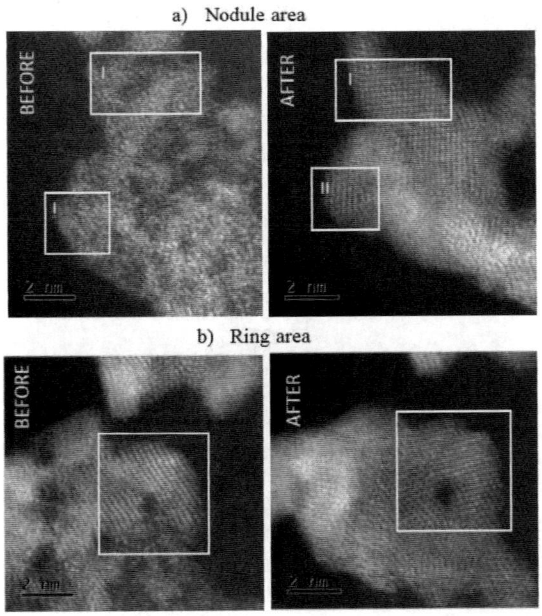

Figure 1 Wide field of view images of the ruthenium catalyst on carbon black, before and after 200kV beam exposure. The analysis regions are highlighted in yellow. The regions were assigned names: a) nodule, with analysis regions I and II and b) ring, with analysis region in yellow.

Scan-drift correction. This correction ensures that the relative positions of the atomic columns we detect and measure lie within the analysis areas. We implemented the non-rigid technique here[2].

Detector asymmetry. Asymmetry of the ADF detector has to be compensated to make reliable correlations of scattering cross-sections to atomic counts using simulations[3]. This was addressed by using a technique proposed by Martinez et al.[4].

Detector current normalisation. Throughout the imaging process, the current the detector measures drops non-linearly due to the electron gun. The measured column intensity would therefore decrease when measured for many images over time. This deficit was compensated, and corrected, in our measurements by measuring the detector current at several imaging intervals.

After these corrections, the resultant frames were compiled into a movie (https://goo.gl/nlIi6A). From this, the scattering cross-sections were measured using absolute integrator software[5]. Once the measured cross-sections were obtained they were then correlated to simulations. Simulations were run using the μSTEM software[6]. The simulation supercell was 2048x2048 pixels, tiled 8times in x and y, then a multislice simulation was carried out using the frozen phonon model with 30 phonons.

3. Correlating cross-sections to atom counts.

The idea of using scattering cross-sections to estimate atomic column counts was first suggested by Retsky[7], and more recently expanded upon by E et al[8]. Scattering cross-sections can be correlated to atomic column counts through simulations. The measured cross-sections are robust to several microscope parameters such as: magnification, defocus, convergence angle, source size, astigmatism, scan noise and sample mis-tilt[8].

Using a technique demonstrated by Jones et al[5] the measured cross-sections were correlated to simulated atom counts. The simulated library for the ruthenium nanoparticles is shown in Figure 2. In some frames the atomic columns were difficult for Absolute Integrator to resolve. This was due to smearing of the intensities during scan-drift correction and the catalyst's phase shifting constantly.

Therefore, only frames where the atomic columns were reasonably well resolved, within the quantification areas, were used. From Figure 2, atomic maps of the quantification regions were created by assigning atoms of simulated cross-sections to closest matching measured cross-sections. A simulation library of fcc [110] was used ([100] and [110] orientations yield similar cross-sections for thicknesses of up to 8 atoms, as shown by Figure 2). From this, three-dimensional surface plots were derived, shown in Figure 4.

Journal of Physics: Conference Series **644** (2015) 012035 doi:10.1088/1742-6596/644/1/012035

Figure 4 displays snapshots of the beginning, intermediary and final evolution phases of the catalyst. The carbon-black support, can be interpreted as the white surface the catalyst disperses on. The resolvable frames can be seen at (https://goo.gl/nlIi6A).

The crystal class of the structures in Figure 4 was determined by super positioning perfect fcc and hcp lattices. Overlaying several images in this manner suggested that the regions observed in this case were fcc 100 and 110 for nodule regions I and II respectively (Figure 3). Whereas, the ring region was a mixture of the hcp and fcc, continuously changing between the two throughout imaging.

Next, the mean thickness of the structure was measured by plotting the mean atomic column counts against imaging time, Figure 5. Figure 5, is a plot from nodule regions I and II. The plot indicates stabilisation of the raft after a certain imaging period. The ring region also yielded similar results.

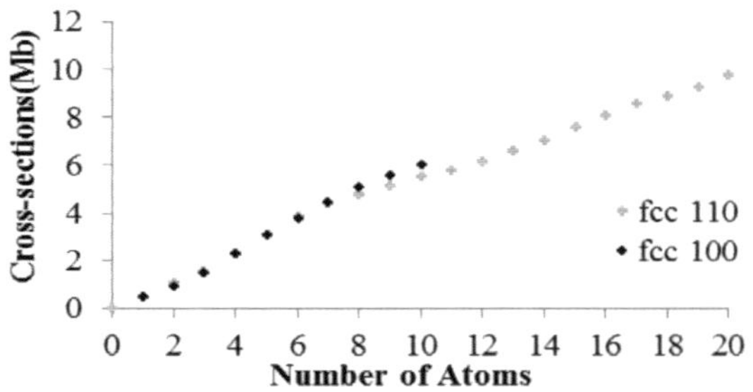

Figure 2 µSTEM simulation library for fcc [100] and [110] used for correlating measured cross-sections to number of atoms within an atomic column. Simulated cross-sections between the two orientations are very similar for thicknesses up to 8 atoms.

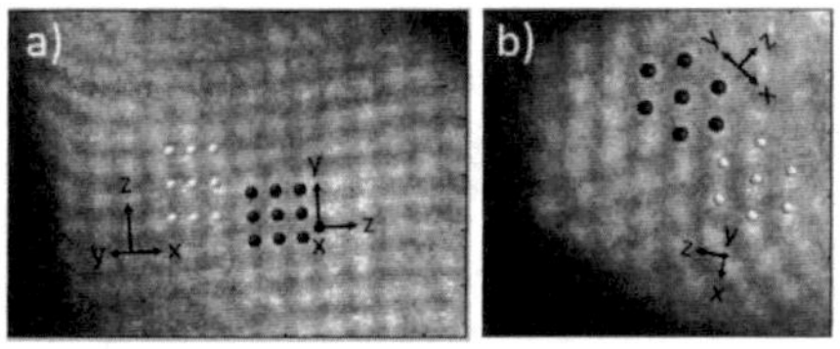

Figure 3 a) and b): Images of regions I and II with perfect fcc (Black) and hcp (Yellow) candidate lattices.

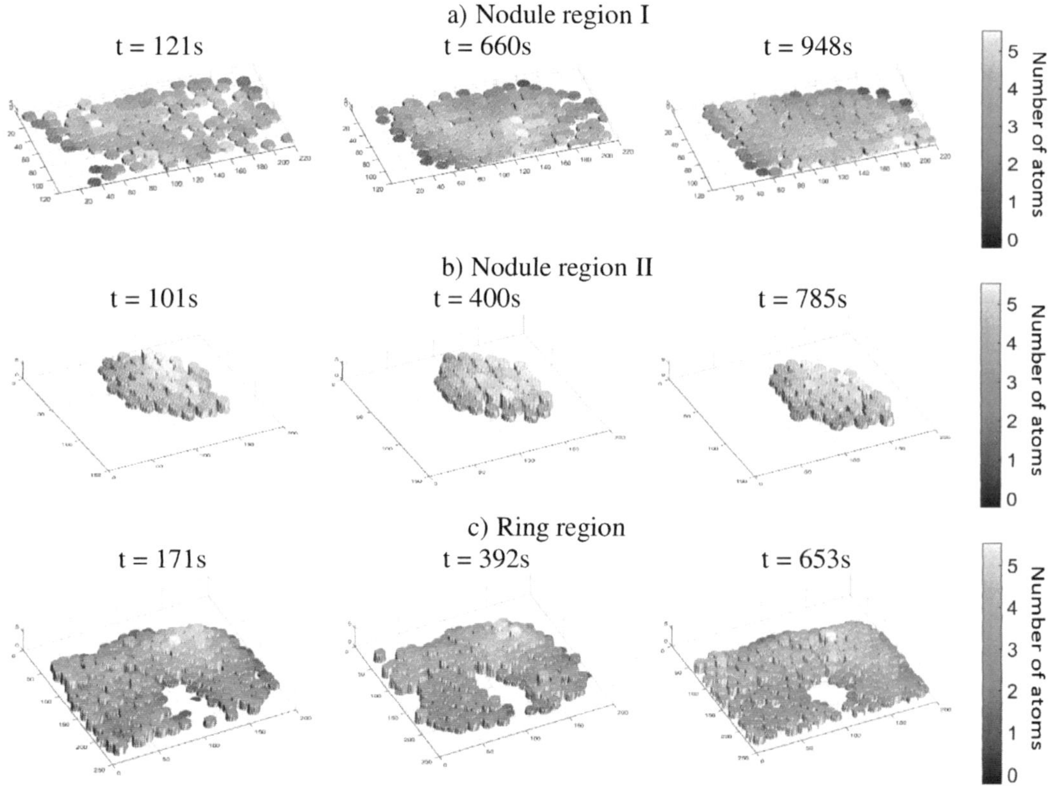

Figure 4 a set of surface plots displaying how the structure changed under beam effects at different beam exposure times, t. The x and y axes represent the pixel coordinates of the atomic columns.

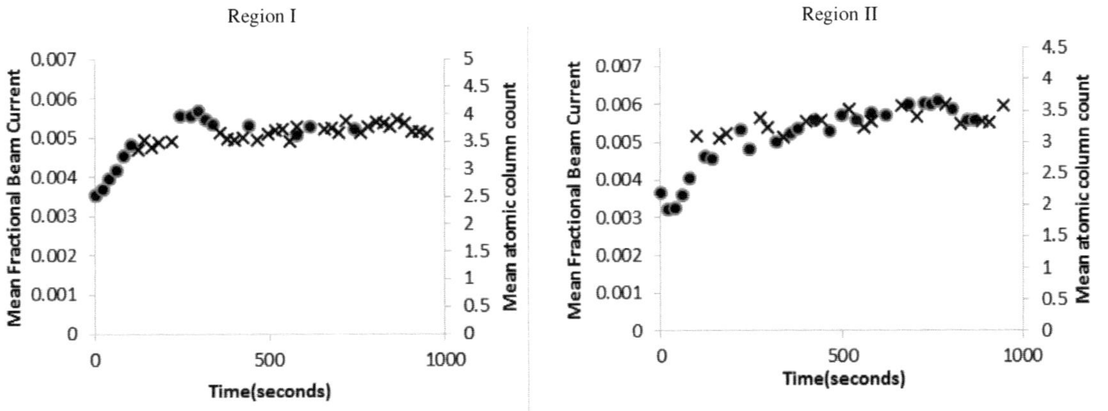

Figure 5 Mean thickness plot for Nodule. The crosses represent counts in atoms whereas the dots represent mean fractional beam current. The mean fractional beam current was assigned to frames that were difficult to resolve. The plots indicated a mean thickness range between 2 and 4 atoms.

4. Discussion and conclusions.

We have presented quantification data of a ruthenium catalyst on carbon-black support as the sample was exposed to the electron beam. Our analysis suggests formation of raft-like structures forming under the electron beam. The crystal class was determined to be fcc for the nodule area (whereas a mixture of hcp and fcc for the ring area), this is particularly interesting as ruthenium is commonly known to possess a hcp structure in the bulk. Furthermore, Density Functional Theory and molecular dynamics calculations by Zhao et al.[9] indicate that it may be possible to obtain a stable fcc ruthenium raft-like structure if a carbon atom is added to the more common hcp ruthenium structure. Therefore, in conclusion the ruthenium atoms wet the carbon support and formed a stable fcc raft between 2 and 6 atoms thickness.

5. Acknowledgements.

This work has been funded by the EPSRC and Johnson Matthey as part of a CASE-Award studentship. The authors would like to thank Johnson Matthey for providing the catalyst sample, and the University of Trondheim for provision of experimental facilities and thank the staff for their help.

6. References.

[1] Kusada K, Kobayashi H, Yamamoto T, Matsumura S, Sumi N and Sato K 2013 *J. Am. Chem. Soc.* **135** 5493
[2] Jones L, Yang H, Pennycook T J, Marshall M S J, Van Aert S, Browning N D, Castell R and Nellist P D 2015 *Adv. Struct. Chem. Imaging* **1** 8
[3] Findlay S D and LeBeau J M 2013 *Ultramicroscopy* **124** 52
[4] Martinez G T, Jones L, De Backer A, Beche A, Verbeeck J Van Aert S and Nellist P D 2015 *Ultramiscroscopy* [in press]
[5] Jones L, MacArthur K E, Fauske V T, Van Helvoort A T J and Nellist P D 2014 *Nano Lett.* **14** 6336
[6] Allen L J, D'Alfonso A J and Findlay S D 2014 *Ultramicroscopy*, **151** 1
[7] Retsky M 1974 *Optik (Stuttg)* **41** 127
[8] E H, MacArthur K E, Pennycook T J, Okunishi E, D'Alfonso A J, Lugg N R Allen L J and Nellist P D 2013 *Ultramicroscopy* **133** 109
[9] Zhao Z, Meng C, Li P, Zhu W, Wang Q, Ma Y, Shen G, Bai L, He H, He D, Yu D, He J Xu B and Tian Y 2014 *Nanoscale* **6** 10370

Characterisation of Mg biodegradable stents produced by magnetron sputtering

N Elmrabet, N Botterill, D M Grant, P D Brown

Division of Materials, Mechanics & Structures, Department of Mechanical, Materials & Manufacturing Engineering, Faculty of Engineering, University of Nottingham, University Park, Nottingham NG7 2RD, UK

E-mail: emxnmel@nottingham.ac.uk

Abstract. Novel Mg-minitubes for biodegradable stent applications have been produced using PVD magnetron sputtering. The minitubes were characterised, as a function of annealing temperature, using a combination of SEM/EDS, XRD and hardness testing. The as-deposited minitubes exhibited columnar grain structures with high levels of porosity. Slight alteration to the crystal structure from columnar to equiaxed grain growth was demonstrated at elevated temperature, along with increased material densification, hardness and corrosion resistance.

1. Introduction

Biomaterials used for stents may be metallic, polymeric or composite structures. Metals, such as stainless steel, titanium alloys and super-elastic shape memory alloys such as NiTi, have received much attention due to their biocompatibility combined with their mechanical properties [1]. However, there is interest to develop biomedical implants such as coronary stents, with a temporary function, using biodegradable materials such as Magnesium-alloys rather than degradable polymeric materials with inferior mechanical properties and degradation profiles [2]. Biodegradable stents are expected to provide a temporary opening into narrowed arterial vessels and then progressively disappear once the vessel remodels. The aim is to address the clinical problems associated with in-stent restenosis and long-term thrombosis, and eliminate the need for prolonged antiplatelet therapy [2]. Mg alloys have shown sufficient cytocompatibility and the amount of Mg^{2+} ions released through degradation processes is negligible compared to the physiologic plasma concentration [3] for stent applications. Here, we report on a study to fabricate Mg narrow walled tubes (Mg minitubes), for stent applications, using magnetron sputtering and an investigation into the effect of post deposition heat treatments.

2. Materials and Methods

Highly pure (99.95%) Mg coatings were deposited using a commercial-scale, closed-field, unbalanced magnetron Physical Vapour Deposition (PVD) coating system (Teer Coatings Ltd. UK). Two diametrically opposed circular magnetrons mounted vertically on the chamber side walls were used with pure (99.95%) Mg targets of diameter 57 mm. Commercially available stainless steel spinal needles of outer diameter 1.2 mm and length ~ 80 mm (Becton Dickinson-BD) were used as substrates, covered by a very thin sacrificial layer. Substrates were mounted on supporting pins held on a twin-axis jig connected to a rotary fixture on the chamber base, thus facilitating even 360° coating of the needle substrate within the plasma fields of the two magnetrons.

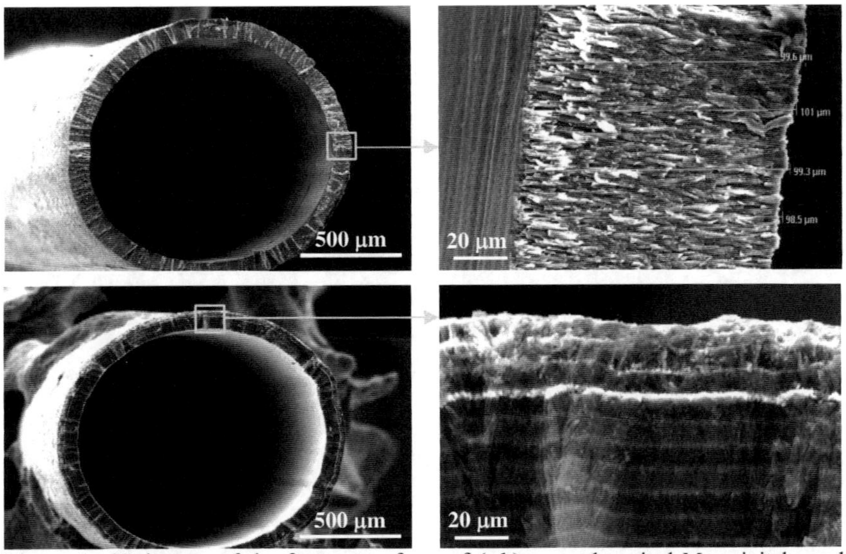

Figure 1 SE images of the fracture surfaces of (a,b) an as-deposited Mg minitube and details of the coating structure; (c,d) Mg minitube annealed at 400°C and details of the annealed coating structure.

A base pressure of 2.0 x 10^{-5} torr was establish prior to deposition and a controlled flow (50 sccm) of Ar (BOC Pureshield, 99.998%) was admitted into the chamber with deposition pressure maintained at ~ 1×10^{-3} torr. Sample rotation was set at 2 rpm, and a DC current of 0.3 A was applied to both Mg targets. Samples were rotated within the deposition plasma for a period of 8 h, which added ~ 9-10 µm of Mg layer thickness. Upon completion of each deposition phase, following a 30 minute cool down period, the chamber was vented, the twin-axis sample stage removed and fresh Mg targets were mounted on both magnetrons; the sample stage was re-admitted to the chamber and pump-down recommenced prior to continued layer deposition. A total of 10 such Mg depositions were conducted in order to build a tube with layer thickness ~ 100 µm. Tubes were coated with a final layer of hydroxyapatite (HA) of ~ 100 nm thickness from a similar sized target of HA by radio frequency magnetron sputtering. The minitubes were heat treated, post-deposition, at temperatures of 300°C, 400°C and 450°C under mixed He/H_2 conditions to eliminate oxidation effects during crystal regrowth. Tubes were also heat treated up to 600°C under Ar to encourage more strongly grain regrowth. Samples were allowed to cool slowly to room temperature under flowing Ar.

The Mg-minitubes were imaged using SEM (FEI XL30, 15 kV, spot size 4, working distance 10 mm), whilst EDS was used to appraise minitube chemistry. Complementary XRD (Bruker AXS D8 Advance, Cu K_{α} (λ= 1.54060 Å), 40 kV, 35 mA) was used to appraise the microstructure of the as-deposited and heat treated minitubes, whilst mechanical properties were appraised using Vickers microhardness indentation in cross-section. A potentiostat (VoltaLab PGZ 100, UK) was used to investigate the corrosion behaviour of the as-deposited and heat treated Mg-minitubes. Corrosion tests were performed using a jacketed three-electrode corrosion cell, with the tested metal cross-section being the working electrode; a $0.5\times10\times10$ mm Pt sheet the counter electrode; and saturated calomel (Hg_2Cl_2) the reference electrode (Thermo-scientific, UK). Phosphate buffer solution (PBS, 125 mL, pH 7.2 ± 0.06) was used as the test solution and a controlled-temperature water bath was used to establish a body temperature of 37±1°C. Prior to sample immersion, the O_2 level in the PBS was reduced by bubbling N_2 at a rate of 2 sccm for 30 min. A 1 h settlement was used to establish the open circuit potential (OCP, E_{oc}). A potential scan rate of 5 mV min^{-1} was applied, from -200 mV to +500 mV with respect to E_{oc}.

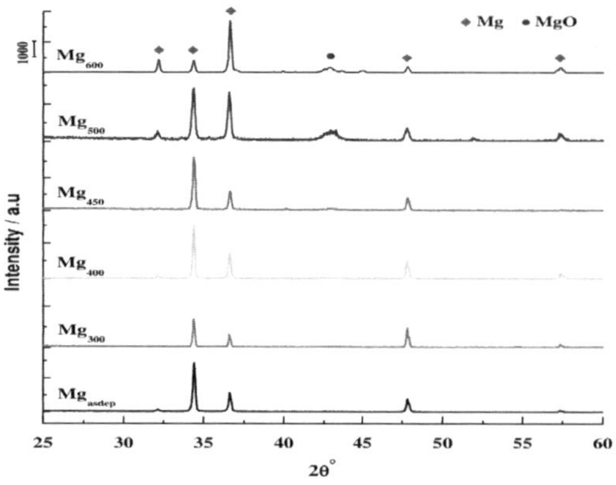

Figure 2. XRD patterns for the as-deposited and annealed Mg-minitubes

Table 1. Hardness values and corrosion parameters for the as-deposited and annealed Mg minitubes

Sample	Hv	Corrosion behaviour	
		E_{cor} (V)	CR (mm/yr)
$Mg_{as\text{-}dep}$	38.7 ± 3.5	-1.66	270.1
Mg_{300}	40.0 ± 6.1	-1.55	45.2
Mg_{400}	51.6 ± 4.1	-1.57	67.1
Mg_{450}	31.9 ± 6.5	-1.55	17.3
Mg_{500}	52.1 ± 9.6	-1.57	4.6
Mg_{600}	32.6 ± 7.3	-1.37	16.0

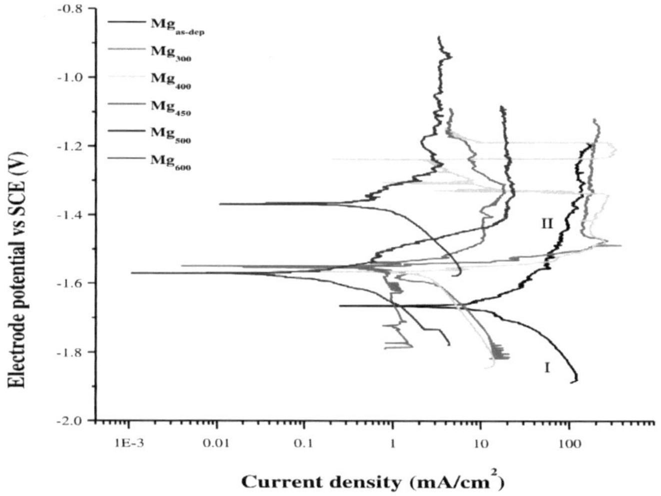

Figure 3. Potentiodynamic polarization curves for as-deposited & annealed Mg-minitubes in PBS at 37°C

3. Results

The Mg minitubes were successfully produced by magnetron sputtering. The total tube wall thickness was ~ 100 μm, as expected. In general, the minitube internal surfaces exhibited wrinkles and folds due to the influence of the sacrificial substrate, whilst the outer surface exhibited rough surfaces with voids and cracks. Figure 1a illustrates the fracture surface of an as-deposited Mg minitube, whilst figure 1b indicates brittle fracture and the presence of a columnar grain structure. Figure 1c illustrates an Mg minitube following annealing at 400°C, whilst figure 1d illustrates details of a layered structure, indicative of oxidation at the interfaces between the sequential Mg deposits, whilst also suggesting a less brittle fracture mechanism. EDS investigations revealed the presence of some Si within the minitubes, originating from the sacrificial layer.

Complementary XRD data (figure 2) confirmed that the as-deposited minitubes were single phase, polycrystalline, columnar-structured Mg, with an intense 002-reflection suggesting preferred growth along the c-axis, which is typical for magnetron sputtered Mg coatings. The as-deposited HA capping layer was found to be amorphous (~100 nm thick). Evidence for MgO formation was detected in samples annealed above 400°C. The average hardness of the as-deposited samples was 38.7±3.5 Hv (Table 1), which is in the lower range for bulk cast Mg, whilst hardness values were variable but increased to a maximum of 52.0 Hv upon annealing at 500°C, suggesting possibly an increase in densification.

Figure 3 present typical potentiodynamic polarization curves for the as-deposited and annealed Mg minitubes, immersed in aerated PBS at 37°C. Table 1 also summarises technical values for the onset and rate of corrosion extracted from these curves. The as-deposited minitube showed normal hydrogen evolution in the cathodic region with the development of a typical Tafel region (I). Fast passivation acted to slow down the generation of Mg^{+} ions during rapid anodic passivation (II). The annealed Mg-minitubes showed generally increased noble behaviour, as compared with the as-deposited sample, demonstrating increased resistance to the onset of corrosion, with the minitube annealed at 600°C showing the most resistance at -1.37 V. A general reduction in the rate of corrosion was also measured as a function of annealing, with best performance shown for the hardest sample annealed at 500°C.

3. Discussion

It is evident that using this novel PVD technique to build up Mg structures, layer by layer, can be an appropriate method for the production of thin, narrow-walled biodegradable stents. However, the as-deposited minitubes exhibited weak mechanical properties, *i.e.* low ductility and flexibility, and a highly brittle nature, whilst increased densification and improved corrosion resistance was demonstrated with annealing.

The developed tube microstructures were compared with the models proposed by Thornton for growing PVD thin films, as a function of temperature and working gas pressure [4]. The Mg minitubes exhibited similar features with zones of tapered grains separated by voids. More dense structures preserving initial nuclei orientations were formed at higher temperatures. However, all of the tubes exhibited high levels of porosity. Preferred orientation along the c-axis is recognised as a strain relief mechanism. Full re-crystallisation of the Mg-minitubes at elevated temperature was not achieved, although some alteration from anisotropic to isotropic grain growth was indicated from XRD, suggesting a process of limited recrystallization at voids and open grain boundaries, consistent with the slight increase in hardness.

The Mg-minitube behaviour varied dramatically when exposed to corrosive environments, with the as-deposited minitube being highly reactive within PBS, limiting its practical use within the biological environment. However, the corrosion resistance of the Mg minitubes improved significantly after heat treatment. The combination of partial grain regrowth, increased densification and oxide layer formation at the Mg layer interfaces, and possible oxide formation around the Mg grains, after annealing at ~ 500 – 600°C, is suggested as the reason for the reduced rate and delayed onset of corrosion.

4. Summary

Mg minitubes, being a candidate system for a biodegradable stent, have been produced using PVD. They exhibited columnar growth structures with open grain boundaries. Slight alteration to the crystal structure of the as-deposited minitubes from columnar to equiaxed grain growth was demonstrated at elevated temperature, with slight increase in material hardness, accompanied by improved corrosion resistance in PBS.

Acknowledgements

The authors are very grateful to Dr. Katy Voisey and Dr. Deen Zhang for their contributory discussions and assistance with corrosion tests. NE also thanks Dr. Xuanli Luo for her assistance with the Mg-minitube heat treatments.

References

[1] Hanawa T 2009 *The Japanese Society for Artificial Organs* **12** 73
[2] Hermawan H, Dubé D and Mantovani D 2010 *Acta Biomaterialia* **6** 1693
[3] Brown D, Lee E, Loh C & Kee S 2009 *Journal of Vascular and Interventional Radiology* **20** 315
[4] Thornton J 1977*Annual Review of Materials Science* **7** 239

Barium Titanate Nanoparticles for Biomarker Applications

O Matar[1], O M Posada[2], N S Hondow[1], C Wälti[3], M Saunders[4], C A Murray[5], R M D Brydson[1], S J Milne[1] and A P Brown[1]

1. Institute for Materials Research, School of Chemical and Process Engineering,*
2. Leeds Insitute of Cardiovascular and Metabolic Medicine, School of Medicine,*
3. Bioelectronics, School of Electronic and Electrical Engineering,*
* University of Leeds, LS2 9JT, UK.
4. Centre for Microscopy, Characterisation and Analysis, The University of Western Australia, Perth, WA, Australia.
5. Diamond Light Source Ltd, Harwell Science and Innovation Campus, Didcot, OX11 0DE, UK.

sm09om@leeds.ac.uk

Abstract. A tetragonal crystal structure is required for barium titanate nanoparticles to exhibit the nonlinear optical effect of second harmonic light generation (SHG) for use as a biomarker when illuminated by a near-infrared source. Here we use synchrotron XRD to elucidate the tetragonal phase of commercially purchased tetragonal, cubic and hydrothermally prepared barium titanate ($BaTiO_3$) nanoparticles by peak fitting with reference patterns. The local phase of individual nanoparticles is determined by STEM electron energy loss spectroscopy (EELS), measuring the core-loss O *K*-edge and the Ti L_3-edge energy separation of the t_{2g}, e_g peaks. The results show a change in energy separation between the t_{2g} and e_g peak from the surface and core of the particles, suggesting an intraparticle phase mixture of the barium titanate nanoparticles. HAADF-STEM and bright field TEM-EDX show cellular uptake of the hydrothermally prepared $BaTiO_3$ nanoparticles, highlighting the potential for application as biomarkers.

1. Introduction

Nanoparticles with a non-centrosymmetric crystal structure exhibit second harmonic generation (SHG) of light when illuminated with a femtosecond pulsed near-infrared laser. Such nanoparticles can be used as optical biomarkers to circumvent drawbacks with fluorescent proteins and semiconductor quantum dots, such as photobleaching and fluorescent intermittency (blinking) [1]. Barium titanate ($BaTiO_3$) with a tetragonal crystal structure is known for its ferroelectric properties due to the net dipole moment caused by the offset titanium atom in the oxygen octahedron. However, the reduction of barium titanate particle size is assumed to stabilise the paraelectric (cubic) phase [2].

The size dependence of the tetragonal/cubic phase transition of $BaTiO_3$ has long been debated, and the favoured size dependent transformation mechanism suggests a tetragonal particle core with a cubic shell that dominates the bulk at sufficiently small (< 100 nm) particle size [3]. A recent study has suggested that ferroelectric dipoles are present in all barium titanate nanoparticles regardless of size, and the density of these is a function of particle morphology; they suggest that spherical nanoparticles have fewer ferroelectric dipoles in comparison to faceted nanoparticles of similar size [4].

Journal of Physics: Conference Series **644** (2015) 012037 doi:10.1088/1742-6596/644/1/012037

The $BaTiO_3$ crystal structure can be investigated by X-Ray or electron diffraction. Synchrotron X-Ray diffraction provides a bulk measurement that suffers from crystallite size line broadening in the diffraction pattern, making phase determination difficult. Resolving the phase of $BaTiO_3$ is important when utilizing the nanoparticles as biomarkers that produce second harmonic generation (SHG). Scanning transmission electron microscopy (STEM) however, can incorporate electron energy loss spectroscopy (EELS), a highly spatially resolved analytical technique that can indirectly determine the local phase of nanoparticles by producing characteristic core-loss edges specific to the phase of $BaTiO_3$. Moon *et al.* have suggested that the Ti L_3-edge in $BaTiO_3$ varies in t_{2g}-e_g separation between 2.36 eV for the tetragonal and 1.94 eV for cubic phases [5]; on this basis they report $BaTiO_3$ nanoparticles exhibit a cubic core and tetragonal shell in contrast to Asiaie *et al.* [3]. Other studies report that a broadening of the leading two peaks at the O *K*-edge of $BaTiO_3$ reflects O 2*p*-Ti 3*d* hybridization related to tetragonal distortion [6]. A further study of cubic $SrTiO_3$ reports that a reduction in t_{2g} and e_g splitting and broadening of the Ti L_3-edge is related to surface reconstruction and distortion from the cubic symmetry of TiO_6 octahedra [7].

This work utilizes synchrotron XRD to measure the bulk phase and STEM-EEL spectroscopy to resolve the local phase of barium titanate nanoparticles. SHG images of the nanoparticles together with TEM images of cellular uptake highlight the feasibility of using $BaTiO_3$ as biomarkers.

2. Experimental

Barium titanate nanoparticles were synthesized via the hydrothermal method at 150 °C for 72 hours [8]. Commercial tetragonal and cubic phase powders were purchased from Sigma Aldrich (Dorset, UK). Synchrotron powder diffraction was conducted at Diamond Light Source (Oxford, UK) Beamline I11 [9], under the supervision of the co-author stated. The wavelength of the incident beam (λ= 0.82563 Å) was calibrated using a high quality Si standard powder (SRM640c). Peak fitting was conducted using X'Pert HighScore plus software using tetragonal ICCD reference file: 04-013-5890 and cubic ICCD: 01-078-4475 model patterns.

TEM was conducted on two electron microscopes; an FEI Tecnai F20 FEG-TEM operating at 200 kV and fitted with a Gatan Orius SC600A CCD camera and an Oxford Instruments X-Max SD EDX detector, and a non-corrected FEI Titan G2 80-200 operated at 200 kV fitted with a Gatan Ultrascan 2k x 2k CCD camera with a Gatan Enfinium ER EEL spectrometer.

SHG images were taken on an Olympus confocal laser scanning microscope (Fluoview FV 1000) modified for multiphoton microscopy. Using a Ti: Sapphire laser source at 200 mW centered at 820 nm. The laser is pulsed at 80 MHz with a pulse width of 250 fs at the sample. The signal is detected in epi-geometry with a band pass filter centered at 410 nm (± 5 nm).

A549 lung epithelial cells were treated with $BaTiO_3$ nanoparticles (100 μg/ml) in serum free media for a 24 hour period and harvested. The samples were then prepared for electron microscopy by fixation, dehydration, resin embedding and thin sectioning for TEM.

3. Results and Discussion

3.1. Synchrotron XRD. The (002/200) peak from the synchrotron data was fitted to estimate the tetragonal phase fraction for all barium titanate samples. The peak fitting was conducted only on the (002/200) peaks due to the presence of minor phases of titanium rich $BaTi_2O_5$ or hydrated potassium hydroxide. The experimental (002/200) peaks (red) are shown in Figure 1 with the peak fitting analysis overlaid (shaded area) and the cross hairs indicate the fitting of the tetragonal (purple) and cubic (blue) reference data.

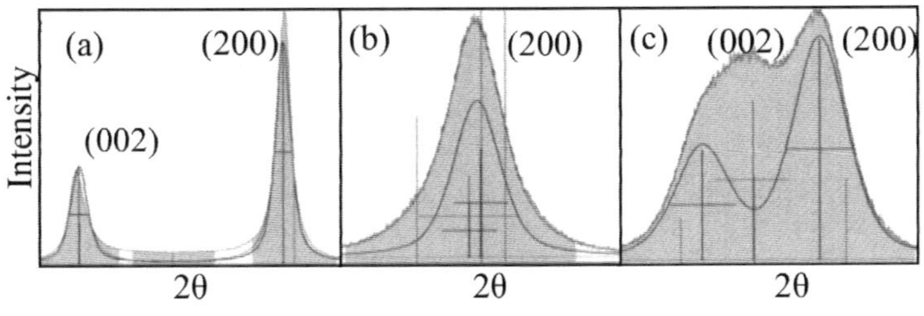

Figure 1. Synchrotron XRD plots of $BaTiO_3$: (a) commercial tetragonal, (b) commercial cubic and (c) hydrothermally prepared $BaTiO_3$.

Peak fitting estimates the tetragonal phase present in (a) commercially purchased tetragonal $BaTiO_3$ to be ~ 94 % tetragonal, (b) commercial cubic $BaTiO_3$ to be ~ 45 % tetragonal and (c) for hydrothermal $BaTiO_3$ synthesised in-house to be ~ 65 % tetragonal.

3.2. Second harmonic generation (SHG). Figure 2 shows that commercial 'tetragonal', 'cubic' and hydrothermal $BaTiO_3$ nanoparticles all produce second harmonic generation of light when illuminated by a near-infrared source. This suggests that the synchrotron XRD analysis is correct in estimating a mixture of cubic and tetragonal phases, as tetragonal is the active phase required for SHG. It is not clear however if this mixture of phases is distributed between particles (*inter*-particle) or within particles (*intra*-particle). Although correlative SHG to SEM imaging suggests the latter (data not shown here).

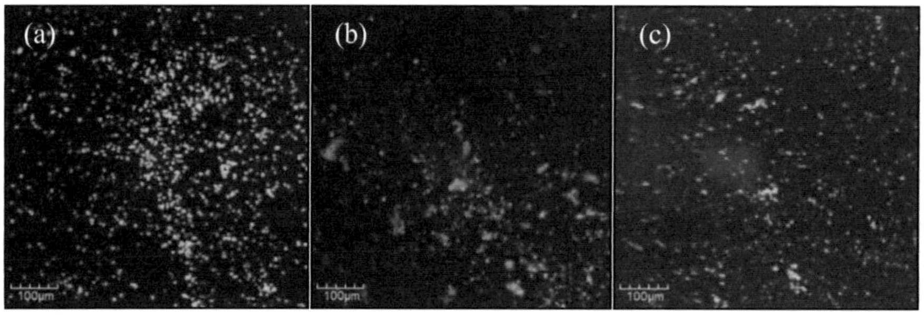

Figure 2. Second harmonic signals of commercially purchased tetragonal (a) and cubic (b) nanoparticles, plus (c) hydrothermally prepared $BaTiO_3$ nanoparticles.

3.3. Scanning transmission electron microscopy and EEL spectroscopy. High angle annular dark field (HAADF)-STEM imaging of $BaTiO_3$ nanocrystals with EEL spectroscopic linescans can indicate the crystallographic phases of $BaTiO_3$ nanoparticles at the particle scale. This technique has been used to investigate the phase content across individual particles at a 2 nm step size. Figure 3 (a) is a HAADF-STEM image of a hydrothermally prepared $BaTiO_3$ nanoparticle. Figure 3 (b) & (c) show the energy separation between the t_{2g}, e_g peaks of the Ti L_3-edges from the core and surface of the particle and the t_{2g}-e_g peak separation of the Ti L_3-edge across the profile of the particle. Figure 3 (d) is the O *K*-edges from the core and surface of the particle.

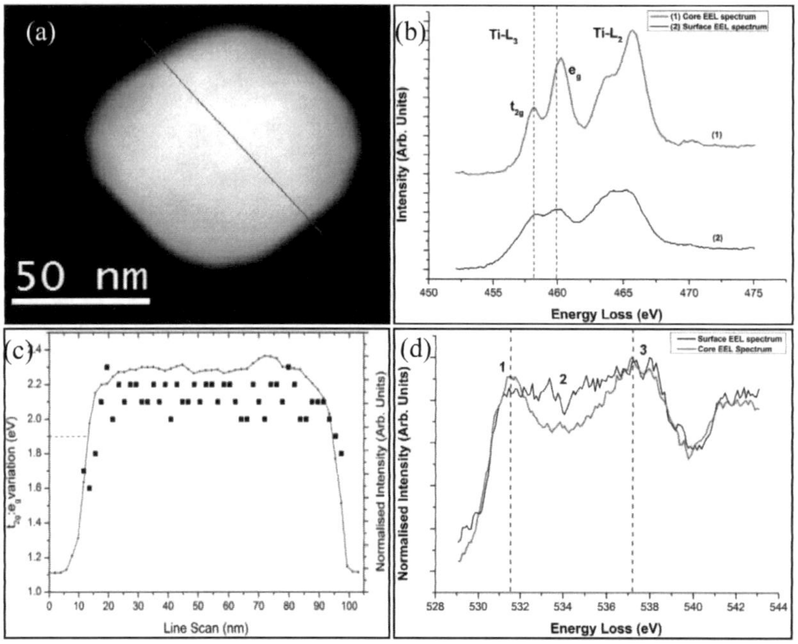

Figure 3. (a) HAADF-STEM image of a hydrothermal $BaTiO_3$ nanoparticle with the line scan location drawn across the particle. (b) Difference between the Ti L_3-edge t_{2g} and e_g peaks at the surface and core of the particle. (c) Energy separation of the Ti L_3-edge t_{2g} and e_g peaks measured across the particle at a 2 nm step size. Dotted line is the HAADF intensity profile. (d) O *K*-edge of particle surface and core. Features (1 & 2) are O 2*p*-Ti 3*d* states split into t_{2g} and e_g, (3) is O 2*p*-Ba 5*d* states.

Figure 3 shows a broadening of both the Ti L_3-edge and the O K-edge at the surface of a $BaTiO_3$ nanoparticle which may indicate distortion of TiO_6 octahedra and could possibly suggest a tetragonal shell and cubic core in agreement with the findings of Moon *et al.* [5]. However the t_{2g}-e_g splitting at the Ti L_3-edge is reduced at the surface, which based on the data of Moon would suggest the opposite (i.e. a tetragonal core and a cubic shell) in agreement with [3] – although in absolute terms the splitting are low for both phases. Further work is in progress; however whichever explanation is correct, the results do indicate that there is an intraparticle variation in structure within the nanoparticles with a core volume fraction of ca. 80-85 % not dissimilar to the findings of synchrotron XRD, i.e. this intraparticle phase/structure variation could account for SHG exhibited by the nanoparticles.

Although the phase of a $BaTiO_3$ nanoparticle is important for second harmonic generation of light, if the nanoparticles are to be used as biomarkers; they need to show uptake by cells; Figure 4.

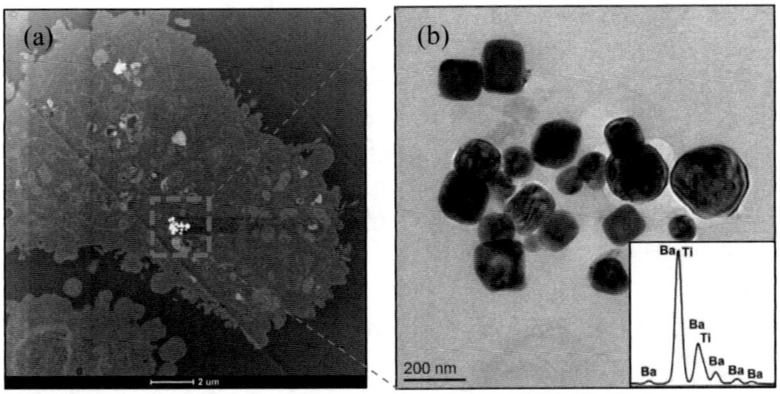

Figure 4. (a) HAADF-STEM image of a thin cell section with an internalized agglomerate of barium titanate nanoparticles. (b) Bright field TEM of the agglomerated nanoparticles from (a), with the associated EDX spectrum revealing barium and titanium inset.

4. Conclusion

Peak fitting of synchrotron XRD suggests that commercial 'tetragonal' and 'cubic', and hydrothermally prepared barium titanate nanoparticles all contain mixtures of tetragonal and cubic phases. Evidence that a tetragonal phase is present in all samples is shown by single and agglomerated particles emitting second harmonic light.

STEM-EELS linescans across $BaTiO_3$ nanoparticles show that there is intraparticle variation in the phase or structure with evidence for some form of surface reconstruction which could account for the phase fractions identified by synchrotron XRD and the presence of SHG.

HAADF-STEM and bright field TEM-EDX show cellular uptake of the hydrothermally prepared $BaTiO_3$ nanoparticles, highlighting the potential for application as biomarkers.

References

[1] Rogov A, Mugnier Y and Bonacina L 2015 *J. Opt* **17** 33001
[2] Frey M H and Payne D A 1996 *Phys. Rev. B* **54** 3158
[3] Asiaie R, Zhu W D, Akbar S and Duta P K 1996 *Chem. Mater* **8** 226
[4] Polking M J, Han M G, Yourdkhani A, Petkov V, Kisielowski C F, Volkov V V, Zhu Y, Caruntu G, Alivisatos P and Ramesh R 2012 *Nat. Mater* **11** 700
[5] Moon S M, Wang X, and Cho N H 2013 *Microsc. Microanal* **19** 123
[6] Bugnet M, Radtke G and Botton G A 2013 *Phys. Rev. B* **88** 201107
[7] Zhu G, Radtke G and Botton G A 2012 *Nature* **490** 384
[8] Eckert J O, Hung-Houston C C, Gersten B L, Lencka M M and Riman R E 1996 *J. Am. Ceram. Soc* **79** 2929
[9] Thompson S P, Parker J E, Potter J, Hill T P, Birt A, Cobb T M, Yuan F and Tang C C 2009 *Rev. Sci. Intrum* **80** 075107

Analysis of Electron Beam Damage of Crystalline Pharmaceutical Materials by Transmission Electron Microscopy

M S'ari [1], J Cattle [1], N Hondow [1], H Blade [2], S Cosgrove [2], R M Brydson [1] and A P Brown [1]

[1] Institute for Materials Research, School of Chemical and Process Engineering, University of Leeds, Leeds, LS2 9JT, United Kingdom
[2] Pharmaceutical Development, AstraZeneca, Macclesfield, SK10 2NA, United Kingdom

E-mail: sm11msa@leeds.ac.uk

Abstract. We have studied the impact of transmission electron microscopy (TEM) and low dose electron diffraction on ten different crystalline pharmaceutical compounds, covering a diverse chemical space and with differing physical properties. The aim was to establish if particular chemical moieties were more susceptible to damage within the electron beam. We have measured crystalline diffraction patterns for each and indexed nine out of ten of them. Characteristic electron dosages are reported for each material, with no apparent correlation between chemical structure and stability within the electron beam. Such low dose electron diffraction protocols are suitable for the study of pharmaceutical compounds.

1. Introduction

The performance of pharmaceuticals is highly dependent on the structure and purity of the active pharmaceutical ingredient (API). It is essential to fully understand and characterize the solid state properties of pharmaceutical materials, different polymorphs (different crystal structures of the same compound) can exhibit different effects in the body and may require different processing and storage conditions. Characterization includes understanding what crystal forms and impurities are present within the material and in what quantity. Currently the conventional ways to determine the properties of materials are to use techniques such as powder X-ray diffraction (pXRD). These techniques are very useful for analyzing bulk materials however, current detection limits make it difficult to analyze trace amounts (<1%), materials with low crystallinity or nano-sized crystalline particles.

Transmission electron microscopy (TEM) can be applied to the characterization of pharmaceuticals, for example, through analysis of individual particles by electron diffraction, the identification of amorphous and crystalline phases present at low concentrations and identification of different polymorphs [1]. The diffraction patterns yield information on the symmetry and unit cell dimensions of the crystalline form within the material, ultimately characterizing the API and distinguish between different crystal structures. Through analysis of particles or crystals of different morphology, trace impurities can be identified and analyzed using TEM; by electron diffraction, energy dispersive X-ray (EDX) spectroscopy and potentially electron energy loss spectroscopy (EELS).

A major problem in TEM analysis of pharmaceuticals is alteration or damage due to the electron beam, especially for structurally sensitive specimens such as biological materials, polymers and organic crystals. There are two main causes of beam damage that occur in TEM, elastic and inelastic scattering events [2, 3].

Elastic scattering occurs when electrons interact with the electrostatic field of atomic nuclei, and can result in atomic displacement or electron-beam sputtering. Inelastic scattering is caused by electron-electron interactions and can result in ionization damage (radiolysis) of the sample, this causes chemical bonds to break and change the structure of the specimen. Other forms of damage that can be caused by both elastic and inelastic scattering include electrostatic charging, hydrocarbon contamination and heating of the specimen [2, 3].

For organic compounds radiolysis is the main form of radiation damage and in some crystalline specimens this results in a decrease in crystallinity, evident in a fading electron diffraction pattern. Previous work on organic compounds has shown conjugated ring systems increase the stability of the compound in the electron beam probably due to improved intermolecular bonding [4]. Replacing hydrogen atoms with halides such as chlorine and fluorine has also been shown to improve the stability of organic compounds possibly due to an increase in steric interactions stopping other atoms from moving once chemical bonds have been broken [5].

In this study we investigated ten different crystalline pharmaceutical compounds covering a diverse chemical space and with differing physical properties to characterise and determine which chemical structures are more susceptible to electron beam damage.

2. Experimental

All the samples were provided by AstraZeneca after checking the crystallinity by pXRD. Samples were prepared for TEM by suspending the non-soluble powders in water and placing a drop onto an Ar/O_2 plasma treated, continuous carbon TEM grid and allowing it to air dry. The grids were examined by TEM (FEI Tecnai F20), operated at 200 kV and equipped with a Gatan Orius CCD camera using a standard room temperature characterization protocol established by Pan et al. [6]. Briefly, the flux of electrons was given by the irradiation time and beam intensity which was controlled by the condenser lens and electron source (extraction voltage, gun lens and spot size) settings.

The characteristic dose (electron dose when the intensity of the brightest diffraction spot falls to a value of 1/e) was obtained from a plot of normalized diffraction spot intensity versus electron fluence. The electron fluence was calculated for each diffraction pattern from the exposure time and electron beam current using a procedure described by Pan et al. [6] and the spot intensities were normalized against the most intense spot.

3. Results and Discussion

Diffraction pattern series were successfully gathered for ten compounds and characteristic dosages determined (Table 1). For the compounds with known structures the experimental diffraction patterns were indexed to a specific zone axis despite not lying in an exact orientation (Figure 1).

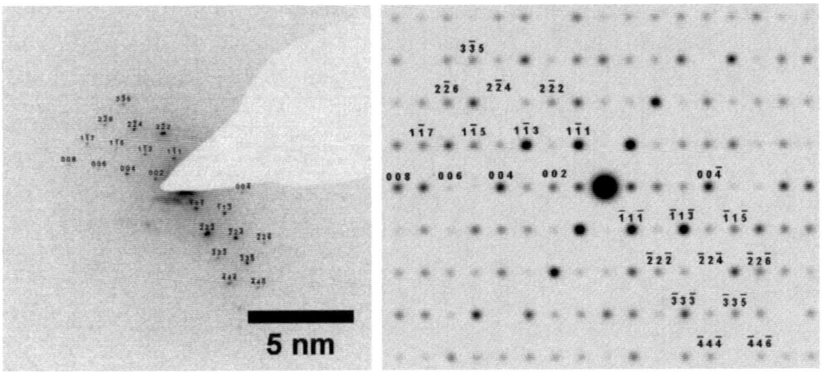

Figure 1: (a) Electron diffraction pattern from an indapamide crystal oriented close to the [110] zone axis, recorded with an electron flux of 3.59 electrons $nm^{-2}\ s^{-1}$ (b) [110] pattern simulation using CrystalMaker software and indapamide structure listed in Table 1.

Journal of Physics: Conference Series **644** (2015) 012038 doi:10.1088/1742-6596/644/1/012038

Table 1: Characteristic dosages (averages and standard deviation of three or more measurements) of specific diffraction spots for each pharmaceutical compound investigated, obtained from normalized intensity versus electron fluence plots (Figure 3).

Compound	Chemical Formula	Characteristic Dose (e^-/nm^2)	Miller Indices (hkl)	CSD Refcode [7]
Tolnaftate	$C_{19}H_{17}NOS$	2079 ± 40	No available structure	No available structure
Griseofulvin	$C_{17}H_{17}ClO_6$	1620 ± 160	011	GRISFL02
Indapamide	$C_{16}H_{16}ClN_3O_3S$	571 ± 40	$\bar{2}22$	FOCCAD
Bicalutamide	$C_{18}H_{14}F_4N_2O_4S$	454 ± 20	002	JAYCES
Cilostazol	$C_{20}H_{27}N_5O_2$	297 ± 85	124	XOSGUH01
Felodipine	$C_{18}H_{19}Cl_2NO_4$	294 ± 40	$\bar{4}11$	DONTIJ
Dutasteride	$C_{27}H_{30}F_6N_2O_2$	220 ± 80	220	LATSIK
Amcinonide	$C_{28}H_{35}FO_7$	130 ± 20	$1\bar{1}1$	VAYJOW
Probucol	$C_{31}H_{48}O_2S_2$	37 ± 10	020	HAXHET01
Efavirenz	$C_{14}H_9ClF_3NO_2$	33 ± 10	211	AJEYAQ02

As mentioned earlier diffraction spots fade as the cumulative dose increases due to the loss of crystallinity within the specimen. In some series, diffraction spots increase in intensity after the initial exposure and then sharply decrease towards zero (Figure 2) this is most probably due to movement/re-orientation of the crystal when irradiated.

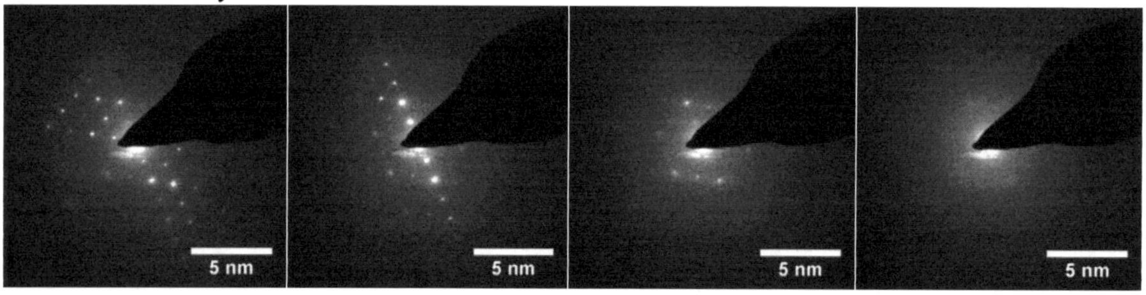

Figure 2: Diffraction pattern series of indapamide initially oriented on the [110] zone axis with electron flux of 3.59 electrons $nm^{-2}\,s^{-1}$ at 45, 120, 170 and 250 seconds after initial exposure to the electron beam. The pattern changes to a systematic row after 120s before fading, consistent with a change in orientation of the crystal as well as a decrease in crystallinity.

Crystallinity decays with a number of different trends as shown by the normalised diffraction spot intensity versus cumulative electron fluence plots (Figure 3). The spot intensities increase and then decrease rapidly (Figure 3a), suggesting re-orientation of a crystal. Figure 3b shows the behaviour from the same sample but uses the ($\bar{2}22$) spot rather than the ($\bar{1}11$) spot. This plot shows a rapid decrease in intensity similar to the previous plot but without the movement/re-orientation. The values of characteristic dose are similar when estimated from each plot. The final trend is shown in both Figure 3c and 3d where there is a decrease in spot intensity followed by a stable plateau and then another decrease. An inverse power-law can be fitted to the final decrease at high fluence in each of the plots in Figure 3, similar to the fitting by Pan et al. who studied the electron beam induced damage of haemosiderin cores containing ferrihydrite [6].

When relating the structures of the pharmaceuticals (Table 1) to the average characteristic doses there are no obvious trends that describe the stability of molecular crystals in the electron beam. This includes characteristic dose and factors such as conjugated rings or halogens, which have been previously shown to increase the stability [4, 5]. Hence, there must be other parameters that contribute to the stability of the organic crystals. Possibilities include the lattice enthalpy (stability of crystal) or enthalpy of sublimation which relates to the intermolecular interactions within the crystal structure [8].

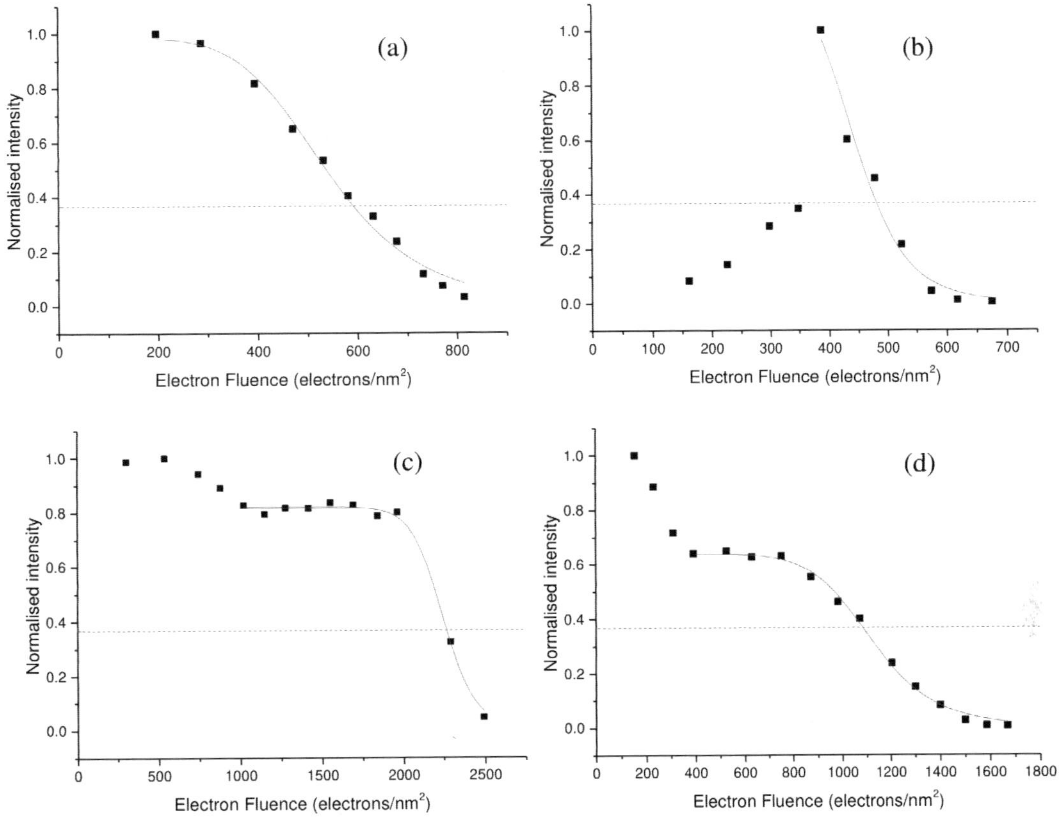

Figure 3: Normalized diffraction spot intensity versus cumulative electron fluence for specific diffraction spots in three different compounds. Red line is the fitting of the inverse power law and the blue line represents the value of 1/e (a) $(\bar{1}11)$ indapamide (b) $(\bar{2}22)$ indapamide (c) tolnaftate, unknown diffraction spot (d) $(\bar{1}11)$ griseofulvin.

4. Conclusion

A low electron dose diffraction protocol for the characterization of pharmaceuticals has been successfully used giving indexed diffraction patterns and initial data on characteristic doses. The next challenge is to extend this protocol to drug formulations. Location of the API in the formulation matrix could also be identified by imaging, diffraction and even spectroscopy (EDX or EELS). Firstly, electron diffraction undertaken within the previously established dose limits can be used to determine the final crystal structure of the API. Secondly, API crystals which are nano-sized, and are therefore difficult to characterise by pXRD, could be characterised using this approach.

References

[1] Eddleston M D, Bithell E G and Jones W 2010 *J. Pharma. Sci.* **99** 4072-4083
[2] Egerton R F 2013 *Ultramicroscopy*.**127**, 100-108
[3] Egerton R F, Li P and Malac M 2004 *Micron* **35** 399-409
[4] Li P and Egerton R F 2004 *Ultramicroscopy* **101** 161-172
[5] Clark W R K, Chapman J N, Macleod A M and Ferrier R P 1980 *Ultramicroscopy* **5** 195-208
[6] Pan Y H, Vaughan G, Brydson R, Bleloch A, Gass M, Sader K and Brown A 2010 *Ultramicroscopy* **110** 1020-1032
[7] Allen F H 2002 *Acta Cryst* **B58** 380-388
[8] Ouvrard C and Mitchell J B O 2003 *Acta Cryst* **B59** 676-685

Bovine Serum Albumin binding to CoCrMo nanoparticles and the influence on dissolution

T A Simoes, A P Brown, S J Milne and R M D Brydson

Institute for Materials Research, SCAPE, University of Leeds, LS2 9JT, UK

E-mail: pmtase@leeds.ac.uk

Abstract. CoCrMo alloys exhibit good mechanical properties, excellent biocompatibility and are widely utilised in orthopaedic joint replacements. Metal-on-metal hip implant degradation leads to the release of metal ions and nanoparticles, which persist through the implant's life and could be a possible cause of health complications. This study correlates preferential binding between proteins and metal alloy nanoparticles to the alloy's corrosion behaviour and the release of metal ions. TEM images show the formation of a protein corona in all particles immersed in albumin containing solutions. Only molybdenum release was significant in these tests, suggesting high dissolution of this element when CoCrMo alloy nanoparticles are produced as wear debris in the presence of serum albumin. The same trend was observed during extended exposure of molybdenum reference nanoparticles to albumin.

1. Introduction

Metal-on-Metal hip implants were intended to offer improved durability, releasing a lower volume of wear debris than traditional implant designs. However the implants may release particles and soluble wear debris that, once in the bloodstream, are surrounded by a diversity of chemical species and hostile environments. Several studies indicate that metal ion levels rise after surgery and could persist. Specifically, metal particles spread by lymphatic circulation can continue to release ions even after removal of the source of wear. A major concern is that these particles and/or ions may trigger an inflammatory response [1]. Most research has concentrated on corrosion/dissolution studies of bulk pieces of CoCrMo hip implants, exploring the influence of possible body environments, in either static or dynamic situations. However, the behaviour of the wear particles after release remains unclear.

It is well known that in a biological environment, the surfaces of biomaterials are modified by adsorption of biomolecules such as proteins, which mediates the interaction between cells and foreign materials. Metal nanoparticles could be fully coated with proteins, effectively becoming a complex biological unit that changes the way in which the body identifies the particles. Depending on the nature of the material, proteins can either enhance or inhibit metal dissolution: a corrosion protection effect is related to protein adsorption on the surface of a material, acting as barrier against corrosive body fluids; whereas complexation of a metal by the amino acids present in proteins could potentially increase corrosion in some materials. It is therefore interesting to investigate whether serum albumin, the most abundant protein found in the circulatory system, may bind to alloy wear debris particles and change the surface, bioavailability and chemical behaviour such as dissolution.

The purpose of this study is to investigate CoCrMo nanoparticle interactions with proteins, specifically isolating the effect of bovine albumin serum. Nanoparticles were generated in large amounts by ball-milling and were comparable in terms of size and composition to some of the nanoparticle wear debris produced by real hip implants [2]. To complement this study, Co_3O_4, Cr_2O_3 and Mo metal nanoparticles were used as reference materials as they may be associated with the surface of the CoCrMo hip implants.

2. Experimental techniques

CoCrMo high carbon gas atomized powders (ASTM75, ~ 60% Co, 27% Cr, 6% Mo) was purchased from Sandvik Osprey. The starting powder had a particle size of around 16 μm. Bovine Serum Albumin (BSA), heat shock, pH 7.0 was acquired from Fisher Scientific, and used during milling as a process control agent. The solution used was composed of 0.4 g of BSA diluted in 10 mL of deionized water. The powder was ball milled for 150 min in a high energy Spex 8000M mill using zirconia ceramic vials and balls. Before milling, the vial was filled with an inert gas (Argon) to avoid atmospheric contamination. A ball-to-powder weight ratio of 3:2 was used. After collection, samples were centrifuged in a Heraeus Megafuge for 10 min, at 25 °C and a centrifugal force of 8000 G. This procedure removes the excess BSA and coarse particles (>200nm), maintaining most of the fine particles in dispersion (20mg/mL). Co_3O_4, Cr_2O_3, Mo nanoparticles (which exhibit MoO_2 on the surface) were purchased from Sigma Aldrich and suspended in BSA (20mg/mL) for 24h, washed in distilled water and finally resuspended in ethanol. The washing procedure removed excess BSA due to its high affinity with ethanol.

In order to demonstrate the effect of milling fluid on the size and stability of CoCrMo nanoparticles when in the presence of serum proteins, the samples were analyzed by Transmission Electron Microscopy (TEM), Energy-Dispersive X-ray spectroscopy (EDX), Inductively Coupled Plasma Mass Spectrometry (ICP-MS) and Zeta Potential (ZP) measurements.

3. Results and Discussion

Wear debris from hip implants was synthetically produced by milling CoCrMo ASTM75 gas atomized powder in Bovine Albumin Serum (BSA). Following milling, the size of the standard ASTM F75 powder decreased to ca. 0.005-1.8 μm (Figure 1). The reduction in particle size is mainly related to friction and impact during milling; however, the interaction with BSA could also play an important role in both the size reduction and the nanoparticle composition. The size fraction of 5-200 nm, commonly found in real wear debris in synovial fluid, was generated here by centrifugal sedimentation (10 min, 25 °C, centrifugal force of 8000 G) prior to testing and analysis.

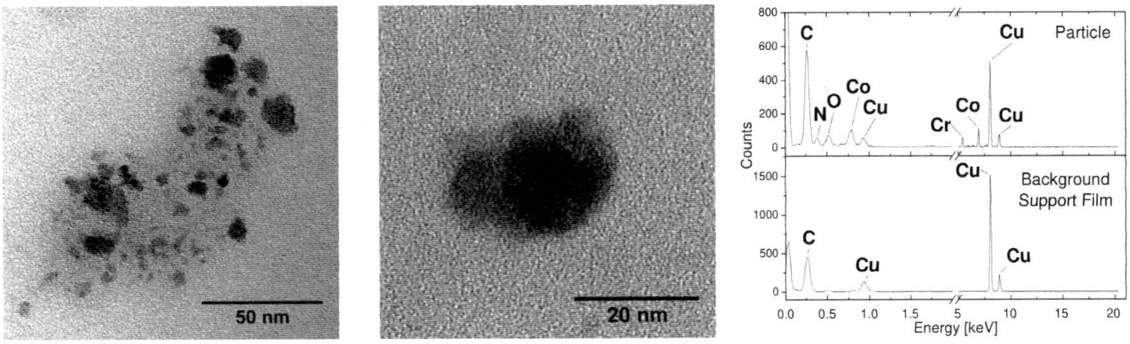

A B C

Figure 1. TEM analysis of synthetically produced CoCrMo nanoparticles on carbon support film; (A) a group of particles surrounded by a thick amorphous layer; (B) and (C) TEM/EDX of a representative particle, showing a composition consisting of cobalt and chromium with the presence of nitrogen and oxygen, note the deficiency in molybdenum signal (Mo $K\alpha$ is at 17.4 keV).

Initial TEM examination shows most of the CoCrMo milled particles were surrounded by a thick amorphous layer (note darker contrast surrounding the group of particles) that could potentially be a protein corona (Figure 1). BSA is formed by a large sequence of amino acids, composed of carboxylic and amine groups, thus the presence of BSA could be distinguished from the TEM carbon support film by the detection of nitrogen. EDX spectra show that particles were mainly composed of cobalt and chromium with a deficiency in molybdenum (relative to the alloy composition), but also reveal the presence of nitrogen and oxygen. A comparison with the composition of the raw material before milling indicates a decrease in cobalt weight percentage in the milled material of only about 5%.

The reference materials showed an average size distribution of 100 nm, with standard deviation of 41nm. The particles where incubated as received for 24 h in BSA, washed and then analyzed by TEM. The results show the clear formation of a protein biofilm (generally less than 5 nm in thickness) on the surface of all the particles (see Figure 2).

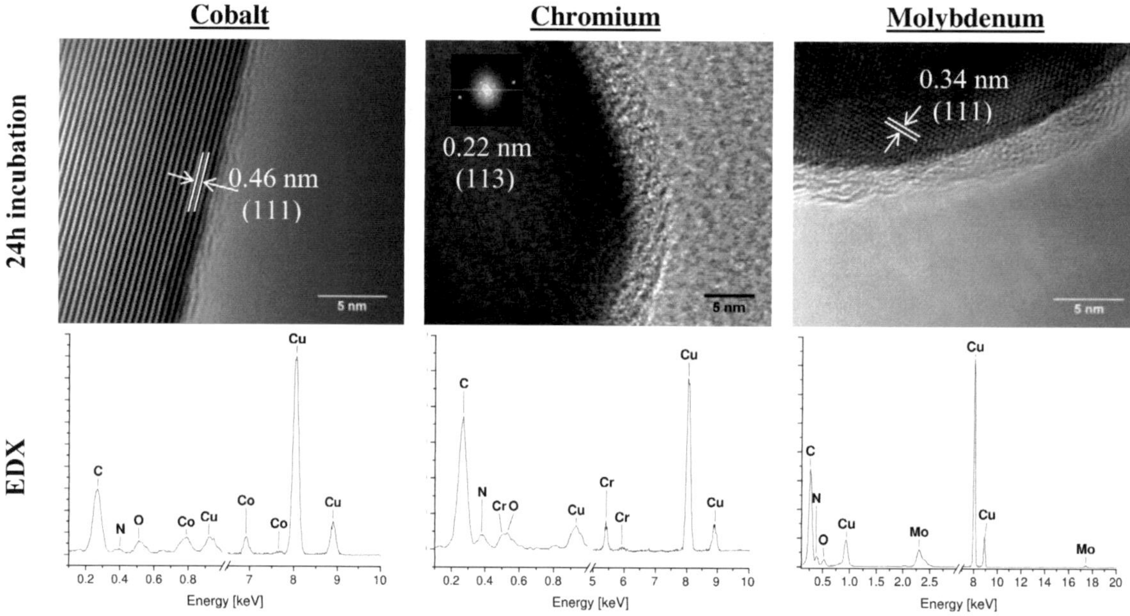

Figure 2. TEM +EDX of Co_3O_4 (JCPDS 42-1467), Cr_2O_3 (JCPDS 38-1479) and MoO_2 (JCPDS 32-0671) reference materials incubated for 24h showing a biolayer containing nitrogen surrounding the particles

Dissolution of the reference materials after one month immersion in a BSA containing solution was measured by ICP-MS (Figure 3) following centrifugation and ultrafiltration to remove any small nanoparticulates [2]. The levels of Co and Cr ions were found to be low, possibly as a result of biolayer formation inhibiting contact with the solution. Molybdenum however, showed high levels of dissolution in the presence of albumin protein in accordance with recent electrochemical corrosion studies of CoCrMo bulk alloy in BSA [4]. Molybdenum is added to the alloy to impart corrosion resistance and its depletion in the presence of BSA, even at pH 7, could leave the material open to greater corrosive attack.

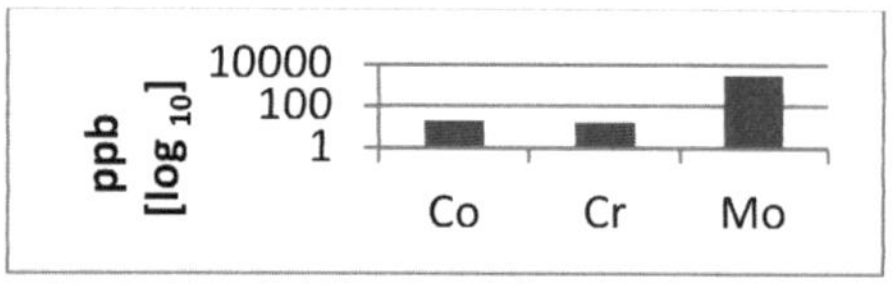

Figure 3. ICP-MS of the dissolution of Co_3O_4, Cr_2O_3 and MoO_2 reference materials incubated for a month in BSA containing stock

Electrostatic forces are expected to play a significant role in protein binding. Albumin, when ionized in water at pH > 4.9 (isoelectric point - IEP), is a negatively charged protein that often shows affinity for hydrophobic surfaces, however it does exhibit a distinct positive patch on one of the triangular protein surfaces [5]. The zeta potential represents the surface charge in the presence of an aqueous solution when functional groups dissociate or ions adsorb onto surfaces from the solution. The IEP is the pH at which a species exhibits a net zeta potential (approximating to surface charge if specifically adsorbed ions are present) equal to zero. At the IEP of a protein, its structure is more hydrophobic, more compact and less stable due to the absence of inter-particle repulsive forces. Zeta potential titration results showed that the standard IEP of the metal oxide reference materials in water all changed to become closer to that of BSA presumably due to the surface coverage (Figure 4). Molybdenum oxide showed the highest affinity to BSA, and cobalt the lowest. The CoCrMo alloy particles milled in BSA also showed a similar IEP to BSA. Electrostatic forces mediate the protein corona formation, molybdenum (oxide) particles are negatively charged at all pH values > 2.5, whereas surface charge of chromium and cobalt oxides is positive below pH ~7 (Figure 4).

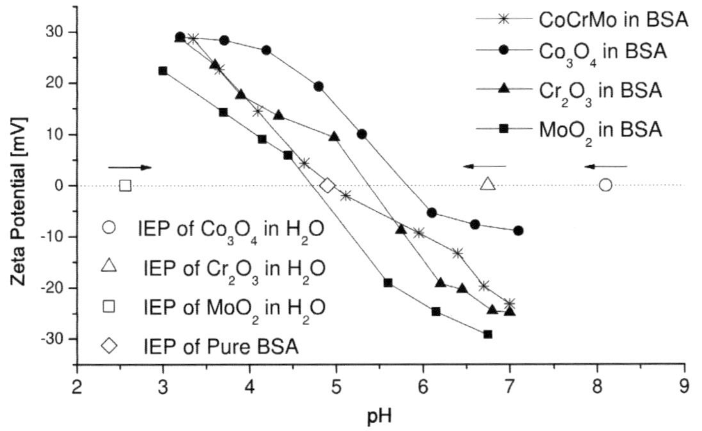

Figure 4. Zeta potential titration of reference materials incubated for 24h in BSA and CoCrMo milled in BSA at the same concentration. The solution pH was titrated from 7 to 3 using HNO_3.

4. Conclusions

TEM analysis suggests the formation of a protein corona on synthetic CoCrMo wear debris particles incubated with BSA and the particles were found to be deficient in Molybdenum by EDX. An interaction with proteins was also seen for elemental reference oxide materials where ICP-MS showed that the reference materials underwent dissolution of predominantly molybdenum ions under static incubation in the presence of BSA. Using the same amount of particles and protein concentration, Molybdenum nanoparticles showed the highest affinity to BSA, demonstrating that in the presence of serum proteins, the effective surface charge is very close to the IEP of BSA. Taken together these results indicate the importance of binding between albumin proteins and molybdenum, which may directly influence the strong dissolution of this element when CoCrMo alloy nanoparticles are produced as hip-implant wear debris in the presence of serum albumin.

Acknowledgements

The authors gratefully acknowledge funding from the Capes Foundation: BEX-5515/10-6.

References

[1] Magone K, Luckenbill D and Goswami T 2015 *Arch Orthop Traum Su* **135** 683–95
[2] Simoes T A, Goode A E, Porter A E, Ryan M P, Milne S J, Brown A P and Brydson R M D 2014 *JPCS* **522** 012059
[3] Simoes T, Bryant M, Brown A, Milne S J, Ryan M, Neville A and Brydson R 2015 Quantitative assessment of metal dissolution from CoCrMo alloys during tribocorrosion *Manuscript submitted for publication*
[4] Espallargas N, Torres C and Muñoz A I 2014 *Wear, "in press"*
[5] Maffre P, Brandholt S, Nienhaus K, Shang L, Parak W J and Nienhaus G U 2014. *Beilstein J Nanotech* **5** 2036–47

A preliminary electron microscopic investigation into the interaction between A$\beta_{1\text{-}42}$ peptide and a novel nanoliposome-coupled retro-inverso peptide inhibitor, developed as a potential treatment for Alzheimer's disease.

M Sherer [1], N J Fullwood [1], M Taylor [1] and D Allsop [1,2]

[1] Division of Biomedical and Life Sciences, Faculty of Health and Medicine, Lancaster University, Lancaster, LA1 4YQ, UK.

d.allsop@lancaster.ac.uk

Abstract. Alzheimer's disease (AD) is a progressive neurodegenerative condition that results in severe cognitive and functional decline in sufferers and for which there are currently no effective treatments to halt or reverse disease progression. AD is the most common form of dementia and age is the major risk factor for this disease. With worldwide population structures changing as increasing number of individuals survive into old age, there is urgent need for novel disease modifying treatments for this condition, which has profound effects upon sufferers in addition to those around them. Some of us have previously developed a peptide inhibitor of A$\beta_{1\text{-}42}$ aggregation (RI-OR2-TAT) that has been shown to reduce A$\beta_{1\text{-}42}$ pathology *in vivo* in mouse models of AD. ~1690 copies of RI-OR2-TAT have been covalently attached to nanoliposome carrier particles forming Peptide Inhibitor NanoParticles (PINPs), and this study investigated the effect of PINPs upon A$\beta_{1\text{-}42}$ aggregation at the molecular level. Our results show that PINPs are able to reduce A$\beta_{1\text{-}42}$ aggregation and do so by binding early (oligomers) and late (fibrillar) stage aggregates. These results highlight the ability of PINPs to disrupt the formation of multiple A$\beta_{1\text{-}42}$ aggregates capable of causing neurotoxicity and thus provide a strong case for PINPs to be carried forward into early stage clinical trials as a novel therapeutic option for the treatment of AD.

1. Introduction

Alzheimer's disease (AD) is a progressive neurodegenerative disease and is the leading cause of dementia [1]. The disease is listed as one of the most common causes of death in many countries, causing numerous symptoms such as impaired memory, cognitive decline, difficulty in

[2] To whom correspondence should be addressed.

communication, and ultimately in the inability to perform activities of daily living. The major risk factor for AD is age, with the late-onset form (responsible for approximately 95% of cases) affecting individuals over the age of 65 [1]. Population structures continue to change as increasing numbers of people are surviving into old age and a result, the need to address the lack of treatments available for diseases of old age such as AD is clear. AD affects, on average, 1 in 9 individuals over the age of 65 and this figure rises to 1 in 3 individuals over the age of 85 [2]. The characteristic pathological features of AD are the formation of extracellular senile plaques and intracellular neurofibrillary tangles. The senile plaques are formed by self-aggregation of the amyloid-β (Aβ) peptide, which is cleaved from a precursor protein – APP – by β- and γ- secretases, whilst the neurofibrillary tangles are formed from hyperphosphorylated Tau protein. Aggregation of the $A\beta_{1\text{-}42}$ peptide primarily occurs via amino acid residues 16-22 which comprise the sequence KLVFFAE [3,4]. Aβ assembles into toxic aggregates of varying sizes, however it is thought that the most neurotoxic forms are aggregates consisting of a small number of peptides, referred to as oligomers [5]. A multitude of evidence exists to suggest that at least some of the neurodegeneration observed in AD is due to the Aβ peptide. Current therapies available to sufferers of AD are aimed at alleviating the symptoms of the disease rather than impacting upon disease progression. Although these treatments have been shown to reduce the rate of cognitive decline in AD sufferers, their lack of effect upon the underlying cause of the disease means that their beneficial effect is only temporary. In order to prevent the progression of AD, disease modifying therapies (DMTs) are required which address the underlying pathological mechanisms of the disease and thus prevent, or maybe even reverse, the effects of AD. As of yet, no DMTs have progressed past phase III trials and, therefore, there are none available for the treatment of AD.

Previously, we have developed a brain-penetrating retro-inverso peptide inhibitor (RI-OR2-TAT, Ac-rGffvlkGrrrrqrrkkrGy-NH2) that blocks the in vitro formation and toxicity of the Aβ oligomers whilst also stimulating neurogenesis [6]. Previous studies have indicated that the aggregation inhibiting activity of this peptide is due to competitive binding between RI-OR2-TAT and $A\beta_{1\text{-}42}$ monomers for the KLVFF region of the $A\beta_{1\text{-}42}$ peptide [4]. In this project RI-OR2-TAT has been attached to the surface of liposomes to attempt to produce a multivalent inhibitor of $A\beta_{1\text{-}42}$ toxicity (termed Peptide Inhibitor NanoParticles – PINPs), and we have analyzed the interaction of the PINPs with $A\beta_{1\text{-}42}$ at the molecular level by transmission electron microscopy (TEM). It is hoped that RI-OR2-TAT will prove to be a successful DMT for AD, thus providing a more positive prognosis for the tens of millions of people suffering from this disease worldwide [7].

2. Materials and Methods

2.1. PINPs
PINPs were produced by Dr M Gregori and Dr M Masserini at the University of Milano-Bicocca, Milan, Italy. Liposomes consist of 1:1 cholesterol:sphingomyelin with 5% maleimide polyethylene glycol (PEG). RI-OR2-TAT is attached to the PEG via the maleimide group by 'click-chemistry' to form PINPs. PINPs were solubilized in either PBS or MilliQ water.

2.2. Transmission Electron Microscopy
Samples were imaged using a JEOL JEM 1010 transmission electron microscope at an operating voltage of 80kV. Samples were placed on 300-mesh carbon/formvar coated copper grids (Agar Scientific, UK) and negatively stained with 2% (w/v) phosphotungstic acid (PTA) pH 7.4.

2.2.1. Immunolabeling
Briefly, 4μl of sample was added to grids for 1 minute, blotted and grids were blocked in 1:10 Goat serum:PBS^+. Samples were then incubated with primary antibody (6E10, BioLegend) at room temperature for 1h, washed with PBS^+, incubated with secondary antibody (G777, Sigma-Aldrich) for

1h, washed in PBS+ and then a further wash in MilliQ water. Grids were then negatively stained by addition 4μl of PTA stain for 1 minute.

3. Results
The purpose of this study was to investigate any interactions between PINPs and $A\beta_{1\text{-}42}$ using TEM.

3.1. Investigating the effect of PINPs upon $A\beta_{1\text{-}42}$ aggregation.

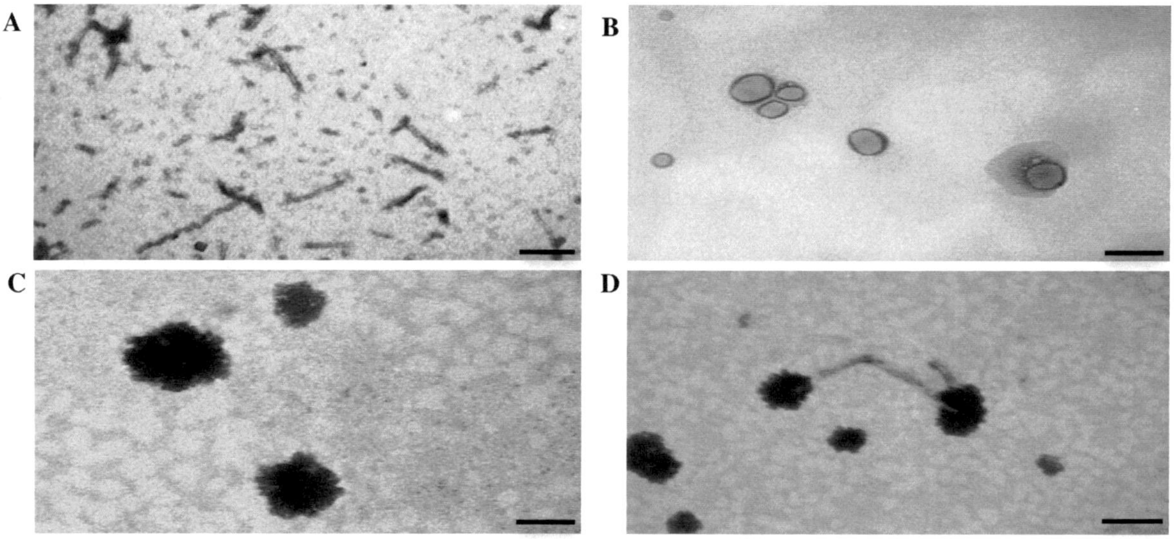

Figure 1 shows the effect of PINPs upon the aggregation of $A\beta_{1\text{-}42}$ following 48-hour incubation at 37°C. (A) $A\beta_{1\text{-}42}$ aggregates in absence of PINPs. (B) PINPs alone. Altered appearance of PINPs (C) and binding of PINPs to $A\beta_{1\text{-}42}$ fibrils (D) following co-incubation. Scale bars = (A) = 200nm (B,C,D) =100nm.

$A\beta_{1\text{-}42}$ samples incubated at 37°C for 48 hours in the absence of PINPs revealed numerous large masses of aggregation, illustrating the tendency of $A\beta_{1\text{-}42}$ to self-aggregate into the multimeric aggregates associated with toxicity in AD (Figure 1A). Interestingly, PINPs + $A\beta_{1\text{-}42}$ coincubation resulted in an altered appearance of PINPs (1B-Before, 1C-After), with PINPs appearing much 'rougher' than when observed in the absence of $A\beta_{1\text{-}42}$. A proposed explanation for this finding was the binding of small oligomers to the PINP surface. The number of $A\beta_{1\text{-}42}$ aggregates observed in the absence of PINPs was increased in comparison to the sample in which PINPs were present. Upon analysis, the $A\beta_{1\text{-}42}$ + PINPs sample clearly showed PINPs binding to $A\beta_{1\text{-}42}$ fibrils of various lengths, with binding observed along the length of the fibril as well as to fibril termini (Figure 1D).

3.2. Immunogold labelling of PINPs to evaluate their ability to bind $A\beta_{1\text{-}42}$ oligomers.

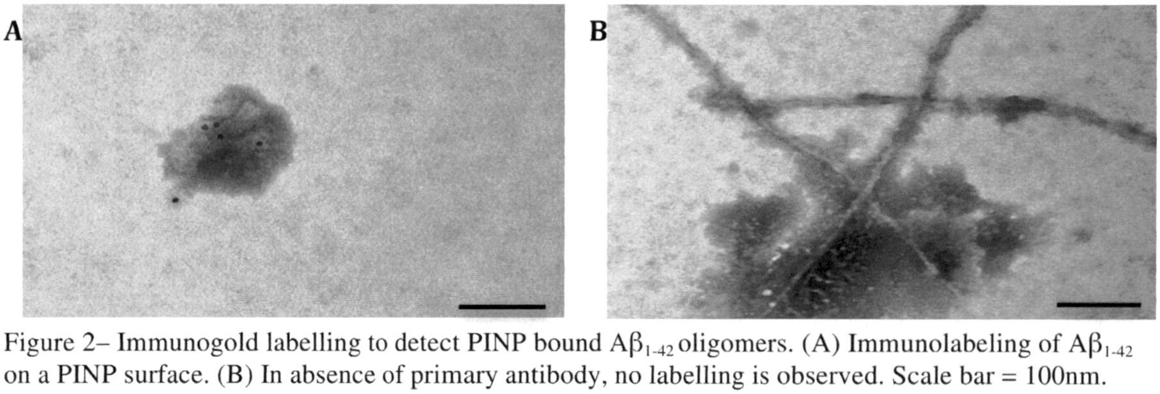

Figure 2– Immunogold labelling to detect PINP bound $A\beta_{1\text{-}42}$ oligomers. (A) Immunolabeling of $A\beta_{1\text{-}42}$ on a PINP surface. (B) In absence of primary antibody, no labelling is observed. Scale bar = 100nm.

To test the hypothesis that the altered appearance of PINPs in the presence of $A\beta_{1\text{-}42}$ was due to binding of $A\beta_{1\text{-}42}$ oligomers, immunolabeling was performed to detect any $A\beta_{1\text{-}42}$ bound to the PINP surface. Results (Figure 2A) showed positive labelling (black dots) on the PINP surface indicating the presence of $A\beta_{1\text{-}42}$ oligomers. Figure 2B shows that in the absence of the primary antibody, no immunolabeling is observed, illustrating that the secondary antibody only binds where the primary antibody is present, showing the specific labelling only where $A\beta_{1\text{-}42}$ is present. Therefore the positive labelling on the PINP surface strongly indicates the presence of $A\beta_{1\text{-}42}$.

4. Discussion

This study investigated the interactions between PINPs and $A\beta_{1\text{-}42}$ at the molecular level in an attempt to evaluate the potential for PINPs to be carried forward as a novel treatment for AD.

Our results indicate that PINPs are able to suppress fibril formation and elongation as co-incubation of $A\beta_{1\text{-}42}$ with PINPs resulted in much fewer dense $A\beta_{1\text{-}42}$ aggregations than was observed with $A\beta_{1\text{-}42}$ alone. Our results show that PINPs are able to bind directly $A\beta_{1\text{-}42}$ fibrils, both along the length of the fibril and to fibril termini. These results suggest that binding of PINPs is able to inhibit further fibril elongation, providing a visual explanation for the termination of fibril elongation reported during analysis of RI-OR2 [4]. Whilst evaluating the ability of PINPs to interact with $A\beta_{1\text{-}42}$, a change in PINP morphology was observed, which we hypothesized was due to the binding of $A\beta_{1\text{-}42}$ oligomers. Binding of $A\beta_{1\text{-}42}$ to the PINP surface was investigated and confirmed using an immunolabeling approach where 10nm gold conjugated antibodies were used to signal positive labeling, demonstrating the ability of PINPs to bind to early (oligomers) and late stage (fibrils) $A\beta_{1\text{-}42}$ aggregates. Thus our results confirm that the peptide has retained the ability to interact with early and late stage aggregates despite attachment to the nanoliposomal carrier particle [4].

This study demonstrates that attachment of RI-OR2-TAT to nanoliposome carrier particles preserves the ability of the peptide to bind oligomers and mature fibrils, reducing total levels of $A\beta_{1\text{-}42}$ aggregation and appearing to inhibit fibril elongation. PINPs were developed as a potential DMT for the treatment of AD. By attaching ~1690 RI-OR2-TAT peptides to the liposome surface, PINPs are able to function as a multivalent inhibitor of $A\beta_{1\text{-}42}$ aggregation, with potential for use in the treatment of AD. We believe that these results demonstrate the strong potential that PINPs possess to be carried forward for further investigation, including early stage clinical trials in order to evaluate their safety and potential therapeutic efficacy in human subjects.

5. Acknowledgements

We would like to thank Maria Gregori and Massimo Masserini from the University of Milano-Bicocca, Milan, Italy who were responsible for the synthesis of the PINPs used throughout this study and without whom this work would not have been possible.

References

[1] Allsop D, and Mayes J 2014 *Essays. Biochem*, ***56***, 99-110.
[2] Hebert L E, Weuve J, Scherr P A, and Evans D A 2013 *Neurology* ***80*** 1778-1783.
[3] Tjernberg L O, Näslund J, Lindqvist F, Johansson J, Karlström A R, Thyberg J, and Nordstedt C 1996 *J. Biol. Chem* ***271*** 8545-8548
[4] Taylor M, Moore S, Mayes J, Parkin E, Beeg M, Canovi M, Gobbi M, Mann D M A, and Allsop D 2010 *Biochemistry* ***49*** 3261-3272
[5] Kim, H, J,. Chae, S, C., Lee D, K., Chromy, B., Lee, S. C., Park, Y, C., Klein, W, L., Krafft, G, A., Hong, S, T., (2003) FASEB J 17: 118–120
[6] Parthsarathy V, McClean P. L, Hölscher C, Taylor M, Tinker C, Jones G, Kolosov O, Salvati E, Gregori M, Masserini M and Allsop D 2013 *Plos One* ***8*** e54769
[7] Reitz C, and Mayeux R 2014 *Biochem. Pharmacol.* ***88*** 640-651.

The effects of Δ^9-Tetrahydrocannabinole treatment on gonadal micro-vascularization and affected fertility examined by SEM and 3D-morphometry

K M T Erlbacher[1] and B Minnich[2]

[1] Institute of Pharmacology and Toxicology, Paracelsus Medical University, Salzburg, Austria
[2] Dept. of Cell Biology, University of Salzburg, Hellbrunnersrt. 34, Salzburg, Austria

E-mail: katharina.erlbacher@sbg.ac.at & bernd.minnich@sbg.ac.at

Abstract. The present study focuses on the effects of Δ^9-tetrahydrocannabinol (THC) on the reproductive system in nude rats with special emphasis on how Δ^9-THC impacts the vascularization of testes which in turn indirectly influences fertility. Basically, Δ^9-tetrahydrocannabinol (THC) causes not only negative (psychoactive) effects in the human body as cannabinole administration in medical use (dose-dependent) offers multiple new treatment opportunities such as pain relief or containment of various cancers. Concerning the reproductive system it strongly influences CB-receptors along the hypothalamic-pituitary-gonadal axis resulting in reduced plasma testosterone levels. There is also altered sperm quality parameters reported such as sperm motility or sperm count. On the other hand Δ^9-THC effects endothelial growth factors (VEGF, Ang-1 etc.) respectively acts on their specific receptors which in turn modify angiogenesis and vascularization of tissues and organs (e.g. tumorous tissues). This leads to new therapeutical strategies in the suppression of various cancers by inhibiting (neo-)vascularization and in turn famishment of tumorous tissues (lack of nutrition supply). Here we studied the micro-vascularization of gonads in a long-term THC-treated nude rat model by vascular corrosion casting, SEM and 3D-morphometry.

1. Introduction

Cannabinoids inhibit tumour angiogenesis in vivo and at least two mechanisms may be involved in this cannabinoid action: direct inhibition of vascular endothelial cell migration and survival as well as the suppression of proangiogenic factors (VEGF, Ang-2) and matrix metalloproteins (MMP) expression in tumours [1]. By inhibiting vascular endothelial cell migration and -survival, cannabinoids would directly prevent blood vessel formation. By targeting tumour cells, cannabinoids would induce their apoptosis [2] and would also suppress proangiogenic factor- and MMP production, further blocking tumour growth and angiogenesis. Cannabinoid-based antiangiogenic treatments thus constitute the most promising antitumoral approaches currently available. Beside these positive angiogenic inhibiting effects, Δ^9-THC may also act in a negative way on gonadal neovascularization and thus may reduce fertility by reduced sperm production and quality. Here we quantitatively studied the micro-vascularization of long-term cannabinole treated nude rats and compared it with plasma serum concentrations of testosterone.

[2] To whom any correspondence should be addressed

2. Material and Methods

2.1 Animals
CR (Charles River) hairless mutation rat (Crl: CD-Prss8) were subdivided into two groups:
(i) Δ^9-Tetrahydrocannabinole (THC) treated adult nude rats (n=30 males), and (ii) healthy control adult nude rats (n=30 males). Animals attending the THC group received 3mg/kg THC i.p. once a day for a period of 182 days.

2.2 Methods
2.2.1 Animal preparation
Adult nude rats were anesthetized using an overdose of Ketavet (130 mg/kg bodyweight diluted in 10 mg/kg Xylazin and 0.9% saline). After anaesthesia the adult animals were pinned with their four ends onto a wax bin. The skin of the animals was opened with a median cut at the dorsum of the abdomen to the dorsum of the episternum and pinned on the side. Next, the abdomen was opened and stretched with a wound retractor. Following, the right *Arteria testicularis* was laid open and ligatured (Ethicon 6-0). After that an injection of 0.1 ml heparin (50 I.E.) and 2-3 minutes later a saturated potassium chloride solution was injected distally of the ligature of the right *Arteria testicularis*. At this point the right *Vena testicularis* was opened.

2.2.2 Vascular Corrosion casting
The vascular system of the testes was rinsed blood-free with phosphate buffered saline (body temperature). Once clear reflux returned from the opened veins the injection was continued with 5-10 ml of the polymerizing resin Mercox (CL-B2-Mix) through the same injection site. At the end of the resin injection the animals were curing about 30 minutes at room temperature and thereafter the cast organs were transferred into a water bath at 60°C over night (tempering) followed by maceration of all organic tissues in potash solution (7.5 %) at 40°C for 2-3 days (changing solution daily). The corroded casts were then rinsed with distilled water and cleaned with a jet of distilled water and HCOOH, thereafter frozen in distilled water and finally freeze-dried. Dry casts were fixed on specimen stubs and sputtered with a thin layer of gold to allow investigation in the SEM [3] (figure1).

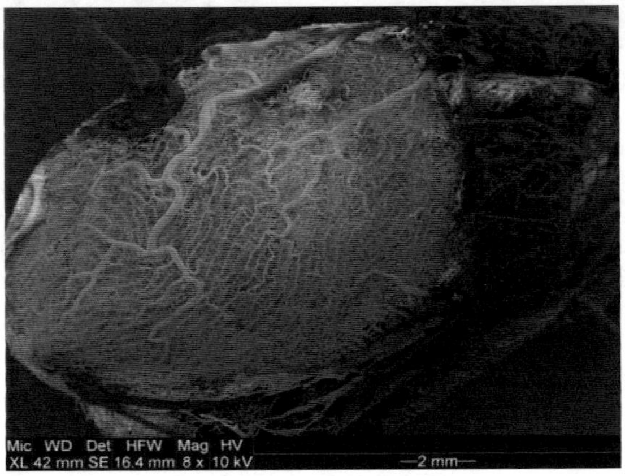

Figure 1. Superficial view of a vascular corrosion cast of a rat's testis examined in the SEM.

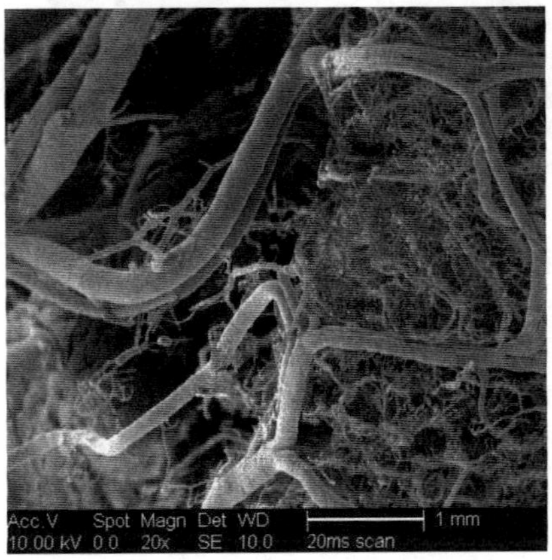

Figure 2. Detailed view of supplying arteries and draining veins in the proximal part of the testis.

2.2.3 SEM inspection
Coated specimens were investigated in the scanning electron microscope (Philips ESEM XL-30, FEI, Eindhoven, NL) at an accelerating voltage of 5-10 kV (figure 2). This accelerating voltage is high enough to identify those details at the (luminal) surface of casted vessels which allow distinguishing

between venous and arterial vessels (e.g. cell nuclei imprints). For consecutive the 3D morphometry analyses stereo-paired images (tilt angle: 5-12°; working distance: 10-30 mm; resolution: 480x480 px) were taken and transferred into M^3 analysis software (ComServ OG, Ebenau, Austria).

2.2.4 3D morphometry

3D representations (figure 3) were generated to gain spatial orientation for point setting. Morphometric measurements were performed considering the third dimension in the SEM by calculating the parallax. Therefore, the type of projection (i.e. parallel projection [magnification (mag) > 500 x and working distance (wd) > 10 mm] respectively, central perspective projection [mag. ≤ 500x and wd ≤ 10 mm]) and the field of view (fw) were critical parameters when acquiring the stereo-paired images (tilt = 6 ± 0.1°). Every point in 3D (x,y,z) was computed from its corresponding 2D-plane co-ordinates (s,t) and (u,v) obtained by point-setting using the computer's input device. For mathematical reasons, we first scaled the planar stereo pair co-ordinates onto a [-1,1] interval within a Cartesian system of co-ordinates and then used homogeneous co-ordinates (x,y,z,1) for computing the necessary transformation matrices and finally applied the appropriate formula (1) or (2) to calculate a point's 3D co-ordinates [4]. With this spatial information geometric vector equations allowed to calculate vascular parameters such as intervascular distances, interbranching distances, vessel diameters and branching-off angles (figure 4).

Figure 3. 3D-morphometric angular measurement – bifurcations of one supplying artery.

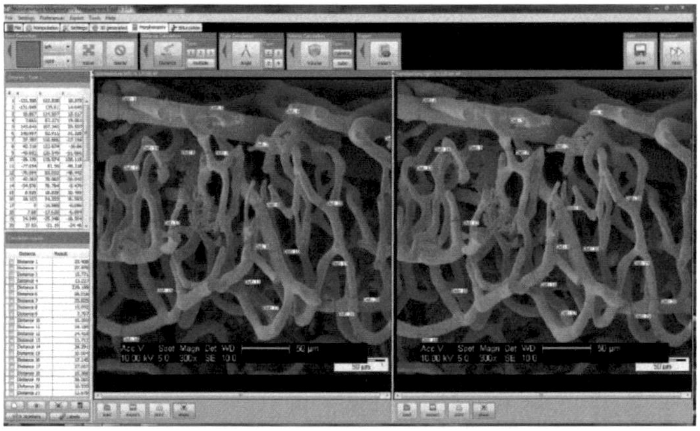

Figure 4. Screenshot of the M^3 morphometry software – vessel diameter measurement on stereo-paired SEM images.

Formulas for 3D point calculation in central perspective projection:

$$x = \frac{10}{N} \cdot (4\, s\, v\, (\sin(\gamma) \sin(^{\rho}/_{2}))^2 + s \sin(2\gamma) \sin(\rho)),$$

$$y = \frac{10}{N} \cdot ((t + v) \sin(\gamma) \sin(\rho)), \qquad (1)$$

$$z = \frac{10}{N} \cdot (4\, t\, v \sin(\gamma)\, (\sin(^{\rho}/_{2}))^2 + (t - v) \cos(\gamma) \sin(\rho)),$$

whereby $N = \sin(2\gamma)\,(1 + \cos(\rho)) + t\, v \sin(2\gamma)\,(1 - \cos(\rho)) + (t - v) \cos(2\gamma) \sin(\rho)$.

Formulas for 3D point calculation in parallel perspective projection:

$$x = \frac{\mu s}{2} \qquad y = \mu \cdot \frac{t+v}{4\cos(\gamma)} \qquad z = \mu \cdot \frac{t-v}{4\sin(\gamma)} \tag{2}$$

3. Results & conclusion

Regarding data in table 1 daily Δ^9-THC treatment over a period of 6.5 month showed a massive change in vascular organization in every part of the testes. Intervascular distances (IVD) the main responsible factor for O_2- and nutrients supply of the surrounding tissue were increased in all cases after THC treatment. With the exception of the medial part of the testicular supply also the diameters of vessels in front of bifurcations were enlarged significantly implicating a higher demand of hormonal supply due to reduced plasma testosterone levels (see figure 5) in THC treated animals. Slightly larger branching-off angles at bifurcations in all parts of the testes in the treatment group indicate a reduced haemodynamic activity which in turn results in a lower blood supply of the sperm producing tissue in the gonads.

In conclusion, we could approve that like in tumors the vascular meshwork is also affected by longtime THC treatment in the gonads which in turn may result in lesser fertility of drug addicted people and cancer patients who receive adjuvant THC therapy.

Table 1. Morphometric comparison of vascular parameters: (a) VD – vessel diameters (b) IVD – intervascular distances, and (c) BA – branching-off angles at bifurcations – controls vs. THC-treated rats' testes divided in distal, medial and proximal portions (mean values).

Segment / Treatment	VD(µm)	IVD(µm)	BA(°)
dist. / control	133.1	213.4	28.2
med. / control	316.0	303.5	36.9
prox. / control	299.7	268.3	39.7
dist. / THC	236.5	315.6	31.8
med. / THC	196.3	581.7	38.1
prox. / THC	421.0	709.1	43.2

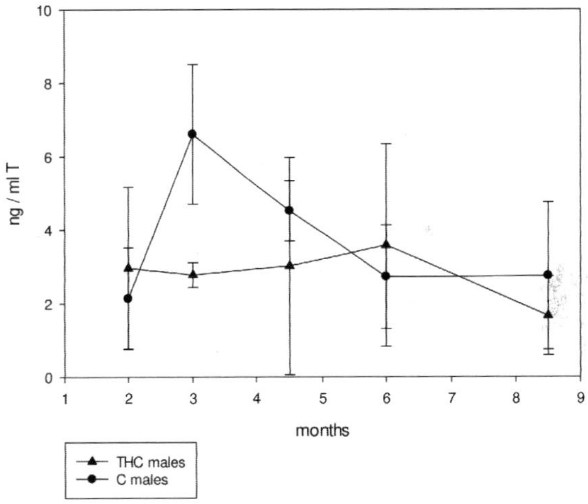

Figure 5. Serum testosterone levels (ng/ml): comparison between THC treated- and control (C) animals.

References

[1] Blazquez C, Casanova L M, Planas A, Gomez del Pulgar T, Villanueva C, Fernandez-Acenero M J, Aragones J, Huffman J W, Jorcano J L, Guzman M 2003 *FASEB J.* **17** 529-31

[2] Sanchez C, de Ceballos M L, Gomez del Pulgar T, Rueda D, Corbacho C, Velasco G, Galve-Roperh I, Huffman J.W., Ramon y Cajal S, Guzman M 2001 *Cancer Res.* **61** 5784-89

[3] Minnich B and Lametschwandtner A 2011 *FORMATEX, Microscopy: Science, Technology, Applications and Education* **I** 29-39

[4] Minnich B, Leeb H, Bernroider E W N and Lametschwandtner A 1999 *J Microscopy* **195** 23-33